All Things Harmless, Useful,
and Ornamental

FLOWS, MIGRATIONS, AND EXCHANGES

Mart A. Stewart and Harriet Ritvo, editors

The Flows, Migrations, and Exchanges series publishes new works of environmental history that explore the cross-border movements of organisms and materials that have shaped the modern world, as well as the varied human attempts to understand, regulate, and manage these movements.

All Things Harmless, Useful, and Ornamental

Environmental Transformation through Species Acclimatization, from Colonial Australia to the World

Pete Minard

The University of North Carolina Press CHAPEL HILL

© 2019 The University of North Carolina Press
All rights reserved
Set in Merope Basic by Westchester Publishing Services
Manufactured in the United States of America

The University of North Carolina Press has been a member of
the Green Press Initiative since 2003.

Library of Congress Cataloging-in-Publication Data
Names: Minard, Peter Maxwell, author.
Title: All things harmless, useful, and ornamental : environmental transformation through species acclimatization, from colonial Australia to the world / Pete Minard.
Other titles: Flows, migrations, and exchanges.
Description: Chapel Hill : University of North Carolina Press, [2019] | Series: Flows, migrations, and exchanges | Includes bibliographical references and index.
Identifiers: LCCN 2019005296 | ISBN 9781469651606 (cloth: alk. paper) | ISBN 9781469651613 (pbk : alk. paper) | ISBN 9781469651620 (ebook)
Subjects: LCSH: Introduced organisms—Australia. | Animal introduction—Australia. | Plant introduction—Australia. | Acclimatization—Australia. | Adaptation (Biology)—Australia.
Classification: LCC QH353 .M55 2019 | DDC 333.95/23—dc23
  LC record available at https://lccn./loc.gov/2019005296

Cover illustrations: Grey peacock-pheasants from Leopold Joseph Franz
Johann Fitzinger, *Bilder-atlas zur Wissenschaftlich-populären Naturgeschichte der Vögel in ihren sämmtlichen Hauptformen* (Vienna: K. K. Hof- und Staatsdruckerei, 1864);
The Acclimatisation Society's Medal (wood engraving) from *The Illustrated Australian News*, 20 June 1868, courtesy of the State Library Victoria.

Contents

Abbreviations in the Text ix

Introduction 1

CHAPTER ONE
Acclimatization Gets Organized 7

CHAPTER TWO
Local Acclimatization Theories 23

CHAPTER THREE
Colonial Creations 43

CHAPTER FOUR
Regulating and Understanding Victorian Fisheries 57

CHAPTER FIVE
Aquaculture 72

CHAPTER SIX
Hunting Victoria 85

CHAPTER SEVEN
The Decline of Terrestrial Acclimatization 108

CHAPTER EIGHT
The Transformation of Fish Acclimatization 121

Epilogue 133

Acknowledgments 137
Appendix: Tables 139
Notes 149   Bibliography 175   Index 193

Illustrations

The late Mr. Edward Wilson 7

Frank Buckland 14

Portrait of a young Frederick McCoy 26

Dr. George Bennett F.Z.S. 30

Ferdinand von Mueller 33

Salmon tanks in Badger Creek 81

Abbreviations in the Text

ASNSW	Acclimatisation Society of New South Wales
ASV	Acclimatisation Society of Victoria
BFAS	Ballarat Fish Acclimatisation Society
FNCV	Field Naturalists Club of Victoria
GWDFAS	Geelong and Western District Fish Acclimatizing Society
IRSAAP	Imperial Russian Society for the Acclimatisation of Animals and Plants
MRFC	Murray River Fishing Company
PIV	Philosophical Institute of Victoria
SAUK	Society for the Acclimatisation of Animals, Birds, Fishes, Insects and Vegetables within the United Kingdom
SZA	Société Zoologique d'Acclimatation
USFC	United States Fisheries Commission
ZASV	Zoological and Acclimatisation Society of Victoria
ZSL	Zoological Society of London
ZSV	Zoological Society of Victoria

All Things Harmless, Useful, and Ornamental

Introduction

Every morning I run a lap of Royal Park in inner-urban Melbourne. On these runs, I often see and hear English sparrows, Indian myna birds, native kookaburras, and magpies. Sparrows and Indian myna birds were introduced to colonial Victoria by the Acclimatisation Society of Victoria (ASV).[1] It sought to legally protect kookaburras and magpies to control pests and so those birds could be exported throughout the British Empire and beyond. Royal Park has been repurposed many times over the last 150 years, housing psychiatric facilities, muster points during both world wars, migrant camps, and emergency housing. Currently it is open parkland that is intended to evoke precolonization Melbourne but also incorporates multiple hospitals, the State Netball and Hockey Centre, and the Royal Melbourne Zoo. The zoo evolved from the ASV and occupies the site granted to it in 1861 for an acclimatization depot.[2]

The depot was a significant nexus in an acclimatization "network of empire" that formed in the 1860s, connecting large parts of the colonized and colonizing world.[3] Reacting to different environments but bound by the idea of rational environmental transformation, acclimatization societies were formed by scientists, dreamers, and landowners throughout Europe, Australasia, and to a lesser extent North America.[4] The French Société Zoologique d'Acclimatation formed first (SZA, 1855), followed by the British Society for the Acclimatisation of Animals, Birds, Fishes, Insects and Vegetables within the United Kingdom (SAUK, 1860).[5] The ASV (1861) was the first acclimatization society formed in Australasia. Its members assisted in the formation of acclimatization societies throughout the Australian colonies and New Zealand, played an active role in organizing acclimatization in Britain, and facilitated the exchange of animals across the globe. The colonial state extensively funded the ASV for several decades.[6]

Money and state support meant that visitors to the Royal Park depot in the 1860s saw a hive of activity. Lakes were dug to support introduced wildfowl and fish; enclosures were created to pen alpacas from Chile, hog and sambar deer from India and Ceylon, and kangaroos and emus ready for export. Aviaries were built, and fruit trees were planted to provide food and shelter for game birds. Some animals, such as the various deer species, were kept at the depot until a sufficient population was acquired and it was apparent they could breed and survive in Victorian climates; they were then released into the wild or sold to private individuals. Other species, such as sparrows, starlings, Indian mynas, and hares, were immediately released at Royal Park, Pentridge Prison, or the nearby University of Melbourne campus and trapped when the ASV wanted to establish populations in more distant parts of the colony.

The invisible and the absent are also important when attempting to understand acclimatization in Victoria. There is much that I cannot see during my morning runs and that many colonists could not easily apprehend when the ASV operated. Despite the ASV's desire, knowledge, money, and connections, the Royal Park depot never contained eland or springbok from Africa, nor was it was possible to successfully ship sufficient gourami from Mauritius to establish a breeding population.[7] These failures are important because they remind us how cosmopolitan and imperial the ASV's vision was, and they warn us not to confuse intent and result. British species became established in Victoria because they were easy to acquire in large numbers, not because of an overwhelming desire to transform the Australian colonies into a facsimile of Britain. Significant experimental plantations were also absent from Royal Park. Botanical experimentation did occur, but at the Royal Botanical Gardens and under the direction of the ASV's vice president, the botanist Ferdinand von Mueller. His forestry and horticultural experiments were conducted autonomously from the ASV and used his separate horticultural network. Acclimatization in Victoria, as conducted by the ASV, unlike in France or even Queensland, became almost exclusively a zoological practice.[8]

A casual visitor to the Royal Park depot would not have been able to observe the complex network of imperial connections, commercial ambitions, travel between colonies, and publications that enabled the ASV to figure out what animals it wanted and to ship animals to and from India, South Africa, Fiji, Mauritius, Ceylon, New Zealand, Sicily, Algeria, France, and the United Kingdom. It is also unlikely that our hypothetical visitor could have ascertained how the ASV's program was shaped by its interactions with farmers, commercial hunters, local scientific societies, anglers, and commercial fishermen. These interactions would lead the ASV to become involved with protecting native and introduced fish, birds, and mammals and discrediting its own terrestrial vertebrate import program within one generation, and the entrenchment of its fish acclimatization activities well into the twentieth century.

An assiduous colonial reader of local newspapers and attendee of public meetings could have perceived how conflicted the ASV and its supporters were concerning the purpose, ambition, and mechanism of acclimatization. This person may also have seen disagreement of the character and potential of Victorian nature—was it primitive and incomplete, rich and full of potential, or damaged and in need of repair? Local men of science used public forums to link acclimatization to Darwinian and or Lamarckian concepts of evolution and constructed understandings of acclimatization that were decidedly antievolutionary.[9] Acclimatization meant many contradictory things to the ASV's various officers and their fellow colonists. Speeches, editorials, and pamphlets discussed acclimatization as a source of beasts of burden for yeoman farmers, food for lost explorers, reminders of Britain and other colonies, opportunities to create and re-create cherished hunting practices, support for a declining commercial fishing industry, control of agricultural pests, and the hope of improving angling in local rivers. Common concerns resonating throughout included how to relate to the imperial center in Britain and other colonies within the empire, what value and values does local nature represent, and how to maintain the colony in the future.

Decline, potential, and transformation are central themes of early colonial Victorian history. The colony began when a small group of speculators from Tasmania landed illegally in Port Phillip Bay in 1835. They negotiated an agreement with some local Wurundjeri people and began grazing sheep on the fertile local plains.[10] Their hope was to make some quick money from pastoralism and land speculation and return to England wealthy men. The reality was not that simple. The imperial and New South Wales governments did not recognize their agreement with the Wurundjeri or their land claims. The colony was allowed to continue, but as the Port Phillip District of New South Wales. During this first period, the colony expanded on the sheep's back. The sheep acted as "shock troops of empire," expanded European dominion, and despoiled Aboriginal pasture land and watercourses.[11] Aboriginal people manipulated, accommodated, imitated, and resisted white colonization to survive.[12] By 1851 Victoria became a self-governing colony; that same year the gold rush broke out, transforming society and damaging local environments. The discovery of gold led to explosive population growth, the creation of phenomenal wealth, and at least some awareness of environmental damage.[13] After the easily accessible alluvial gold was exhausted, many people wanted to break up the large squatter estates and establish intensive yeoman agriculture. The combination of wealth, interest in land reform, and intensive agriculture made Victoria fertile ground for the acclimatization movement. All these resources had their downside. It meant the ASV was able to initiate some real environmental disasters, including the spread of sparrows and Indian mynas. It encouraged and approved of the spread of rabbits but was not itself directly involved in their introduction.

The unintended consequences of acclimatization dominated how it was memorialized for a century or more.[14] Australian scientists and authors, as typified by the mid-twentieth-century zoologist Jock Marshall, argued that the movement represents colonial alienation from the Australian environment: "The bush, to our great-grandfathers was the enemy: it brooded sombrely outside their brave and often pathetic little attempts at civilization; it crowded in on them in times of drought and flood. It, not they, was alien."[15]

To Marshall, this aesthetic disregard and shortsightedness caused Australians to destroy their greatest inheritance, the continent's unique plants and animals. This sentiment is echoed in much of the early scholarly discussions of acclimatization.[16] Harriet Ritvo recently categorized the early verdicts on acclimatization as "a somewhat naive and crude expression of the motives that underlay nineteenth-century imperialism—intellectual and scientific, as well as political and military," that overlook how acclimatization demonstrated the limits of imperial control over nature.[17] Much of the early scholarship discusses the intent and consequences of acclimatization and skips over acclimatization as an evolving practice shaped by many local and global forces. Fortunately, thirty years of scholarship on acclimatization, science, empire, and conservation have helped bridge this gulf of incomprehension. Furthermore, new scholarship on Australasian acclimatization practices has emerged that argues that acclimatizers generated distinct scientific theories to justify acclimatization based on the desire to create a liberal society of small farmers that would use new plants and animals from all over the world to correct colonial environmental damage.[18]

To further develop this scholarship, it is useful to draw on recent developments in environmental history that reexamine the important concept of ecological imperialism.[19] Beattie's acclimatization work is particularly useful when conceptualizing how Australian acclimatizers operated and understood their mission. He invited scholars to think of ecological imperialism not as a one-way transformation of the new world into the old but as a multipolar "empire of the rhododendron" exchanging organisms between colonies and imperial metropolises.[20] The concept of neo-ecological imperialism is very useful when attempting to understand acclimatization. It is the ideas and imperial practices of making colonies profitable after earlier stages of colonization had degraded local environments. Organisms, nutrients, and ideas from elsewhere are exchanged to maintain the profitability and ecological viability of established colonies.[21] This form of ecological imperialism was dependent on complex relationships between colonies. Focusing on multipolar relationships draws heavily on what is still unironically called, after twenty-five years of existence, "new imperial history."[22] This school of thought has collectively suggested that imperial networks are "contested, unstable," mutually constitutive, "webbed multi-centred," and contain momentary connections that come into focus briefly "like the patterns in a kaleidoscope."[23]

These patterns could be investigated in different ways. Some possibilities include attempting to track all the animals acclimatized in Victoria, a simple comparison between Victoria and Britain or France, and a narrative history of the ASV. Instead a hybrid chronological/thematic approach will be attempted that explores acclimatization as an imperial network, social practice, and science—managed in Victoria by the ASV but not entirely controlled by it. The network will be explored by investigating key personnel, events, and introductions undertaken by the ASV and including actors from outside the acclimatization societies, for example, hunters and farmers. This approach allows for exploration of the continuing negotiation of the narratives of acting within acclimatization in Victoria. It will situate acclimatization among traditions that emphasize science, aesthetics, and politics and demonstrate the central importance of locality in the transnational acclimatization movement. Victorian acclimatization was constantly shaped by and shaped local environments, the British Empire, and aesthetics and social aspirations. Victorian acclimatizers were reacting to environmental change, seeking change and restoration, and, ultimately, coming to grips with the consequences of the changes they wrought. Their actions shaped the hybrid environments that I experience on my daily runs and the lives of multitudes of people and animals across four continents.

To make sense of these multitudes and capture ever-shifting and contested acclimatization practices in Victoria, I have adopted the following structure. Chapter 1 explores the establishment of acclimatization in Victoria and the role of Edward Wilson within it, chapter 2 looks at scientific understandings of acclimatization that were established and used in the colony, and chapter 3 explores terrestrial vertebrates imported to and exported from Victoria. Chapter 4 and chapter 5 look at the ASV's understandings of and attempts to regulate colonial fisheries and the ASV's aquaculture program. In chapter 6, I investigate the ASV's contradictory and challenging relationship with hunters and imperial hunting practices. The final two chapters, chapter 7 and chapter 8, explore the next generation of Victorian acclimatizers. They investigate how new ASV members, farmers, and politicians understood and theorized the consequences of earlier acclimatization experiments and largely shut down the importation of terrestrial vertebrates. They also interrogate how new organizations expanded and justified fish acclimatization among a scientific and political climate that was largely skeptical of reckless acclimatization experiments.

CHAPTER ONE

Acclimatization Gets Organized

The late Mr. Edward Wilson. *Australasian Sketcher*, February 16, 1878. Courtesy of the State Library Victoria.

When in 1859 Edward Wilson sighted the "white cliffs of Old England" for the first time in twenty years, he felt underwhelmed and surprised at his lack of patriotic feeling.[1] He feared "that such of my sympathies as are not Australian are cosmopolitan, and not exclusively British. The land of birth is a matter of accident—the land of adoption is a matter of deliberate selection; and having adopted Victoria before her whole population amounted to what she now crams into a single street, and having watched her growth day by day from that time to this, she is the wife of my mature years."[2]

Arriving in Victoria in 1842, Wilson bore witness to an emerging society convulsing through multiple rapid transformations. First he saw the establishment of pastoralism and the displacement of the Kulin nation; then he was both exhilarated and baffled by the gold rush. He participated in the successful colonial separation movement that split Victoria from New South Wales and established responsible government, and sympathized with miners rebelling against unjust taxation during the Eureka Stockade.[3]

By the time of his 1859 visit to England, Wilson was a successful newspaper proprietor (he owned the *Argus*) who was worried about the disruptive effect of manhood suffrage on society, and he was an ardent land reform advocate. His immediate aims in England were to get his cataracts treated and to persuade the British government, landed proprietors, and scientists to aid in the exchange of animals between the Australian colonies and Britain. In 1861 he returned to Victoria and drove the formation of the Acclimatisation Society of Victoria (ASV), served as its founding president, and helped establish a cosmopolitan "network of empire" dedicated to environmental transformations in both Britain and its Australian colonies.[4]

Seeing acclimatization as a network of empire allows for exploration of the mutually constructed and decentered nature of Wilson's influence on the formation of acclimatization in the British Empire. He inflected upon this network a strong interest in the aesthetics and economics of agricultural reform. Wilson, however, had only minimal influence on the science of acclimatization, acclimatization as conservation, or acclimatization for recreation. These matters emerged from the interests of other ASV members, environmental circumstances, and the ASV's increasing involvement with drafting and enforcing game and fisheries acts. However, because of his role in organizing acclimatization in Britain and its colonies, studying Wilson remains critical. Looking at him enables exploration of how, and if, the older acclimatization scholarship and the newer work that discusses environmental transformation and intercolonial relationships can be reconciled.

IN 1857 WILSON COMPOSED a series of letters to the *Argus* maintaining that manhood suffrage and equal electoral districts would hand power permanently to landless workers.[5] Wilson, like many people in the mid-nineteenth century, believed in the progressive development of society over time from hunter-gathering to pastoralism, farming, and then industrial production.[6]

Following from this, he concluded that just as a limited franchise empowered the squatters to retard the development of society by preventing its growth from pastoralism to fully developed agriculture, manhood suffrage would empower workers to skip over developing agriculture and solely focus on protecting their class interests.[7] The 1850s gold rush and subsequent democratic reform shaped Wilson's interests in land reform. Faced with squatters who monopolized land but being suspicious of radical democrats, Wilson maintained his interest in land reform. He believed it might simultaneously act as a bulwark against the political power of the working class and as a way of creating an independent yeoman class that would break the squatters' monopoly on political power. Wilson saw the encouragement of farming and the concomitant proliferation of edible and useful animals within the landscape as essential to the development of the colony and acclimatization as the best means to spur development.

This insight was far from Wilson's alone. In 1855 the Melbourne Chamber of Commerce commissioned a report on developing the colony's agricultural potential.[8] The following year the medical doctor, parliamentarian, and later ASV council member Thomas Embling tabled a parliamentary report recommending importing alpacas into Australia. It was during this debate that Wilson first articulated his support for the systematic introduction of exotic animals, arguing that alpacas could help bring "barren and profitless wastes" under cultivation.[9]

In 1857 Wilson published papers and conducted experiments that indicate a growing fascination with acclimatization's social and political potential. He presented his papers to the Philosophical Institute of Victoria (PIV), the first organized scientific organization in Victoria.[10] The PIV had a strong interest in understanding, exploiting, and managing Victoria's natural resources. By presenting papers to this organization, Wilson was trying to position acclimatization within this practical scientific context. In his first PIV paper, read before the institute in 1857, he made multiple references to Australia and England's economic and social debt to previous unplanned instances of acclimatization. He also noted how Victoria benefited economically from the introduction of sheep, cattle, dogs, cats, goats, and pigs. "How few of the present productions of the colony," he argued, "upon which we are mainly dependent for our comfort and enjoyment were placed here naturally, and without the special intervention of man." "And yet," he continued, "how astonishingly successful their introduction has been!"[11] His papers indicate that he was aware of the role of the introduction of exotic animals in colonization and wished to harness and direct this process via applied zoological science. Wilson's version of acclimatization was as a form of self-aware ecological imperialism.

Wilson's first acclimatization experiment attempted to engineer the ecosystem for the benefit of the empire by translocating native animals within the colony. The experiment consisted of transporting sixty-six Murray cod from King Parrot Creek to the Plenty River near Heidelberg, in the hope that the cod would swim downstream into the Yarra River. Wilson felt that he could provide Victorian anglers, who until then had been limited to catching small native fish, with better game fish, while illustrating the ease and practicality of organized acclimatization.[12] This experiment suggests that for Wilson, acclimatization within the colony was, from its inception, concerned with combining useful and valuable native and exotic organisms, not re-creating England. Wilson's experiments were received with great interest and covered in the daily press, indicating that there was a popular audience for the idea of engineering the local ecology for human benefit. Inspired by Wilson's experiments, attempts were made to establish Murray cod in Lake Colac near Geelong and Lake Burrumbeet near Ballarat.[13]

After experimenting with Murray cod translocation, Wilson attempted to persuade the PIV to begin introducing British songbirds into Victoria. Wilson argued that the systematic and controlled release of one or two species of British birds in large numbers, possibly skylarks or nightingales, would demonstrate that the acclimatization of English birds was possible in Victoria.[14] Wilson explicitly referred to his proposal as an experiment and suggested that only climatically appropriate species, as dictated by latitude, be introduced, positioning his proposal within a Humboldtian understanding of biogeography and practical colonial science.[15] As further proof of the viability of his plan, Wilson reported that English birds had been introduced already near Geelong and at the botanical gardens. Wilson's enthusiasm for establishing British songbirds was a result of the aesthetic and social elements of Wilson's land reform agenda. Songbirds served as a metonym for the imagined cultivated agricultural landscape that Wilson valued so highly. By introducing British songbirds, Wilson was embedding English agrarian imagery in the alien Victorian bush and helping to create a stable cultural landscape. The PIV heartily endorsed Wilson's plan and created a subcommittee dedicated to importing British birds to Australia. Both Wilson and the PIV saw songbirds as a practical means of controlling insect pests and symbolic encouragement to productive, intensive agriculture. These activities were an attempt to use English agrarian imagery to argue for a new improved England in Australia.

Wilson's views on acclimatization and land reform cohered further in 1858 when he left Melbourne to go to England for treatment for his failing eyesight via a convoluted route through New Zealand, South Australia, Sri Lanka, Suez, and Marseille. At the start of Wilson's trip, he visited the colony of South Australia and was inspired by the intensely cultivated, small agricultural holdings he saw there. South Australia was to Wilson "the ecstasy of a realised day-dream." "It is England in miniature—England without its poverty, without its monstrous anomalies of individual wealth-extravagances, thrown into unnecessary and indecent relief by abounding destitution. It is England; with a finer climate, with a virgin soil, with freedom from antiquated abuses, with more liberal institutions, with a happier people; and this is what I have always thought and hoped Australia would become."[16]

Here Wilson articulated the Chartist radical view of land reform. Through taking control of political instrumentalities, working people could create more "liberal institutions," preventing "antiquated abuses" such as tithing, and exorbitant land rents, leading to "happier people." Chartism was an early-nineteenth-century working-class reform movement that aimed to introduce manhood suffrage and parliamentary reform to address the social and political problems of the urban working classes. Chartism also contained a robust agrarian current within its ideology, maintaining that political reform would destroy the gentry's landholding monopoly and enable workers to own small farms and increase their economic independence. After the Chartist movement failed in 1848, Wilson, like many people influenced by Chartism, saw land reform as a way workers could gain some economic independence.[17] Concurrent with South Australia's appeal to his more radical side, Wilson also admired the conservative Wakefieldian elements of South Australia's land system. The keys to Edward Wakefield's theories and the South Australian land system were controlled emigration, the sale of land for intensive agriculture, and colonial self-government.[18] Wilson, like Wakefield, wanted to find the right balance between social dynamism and social stratification through land reform.

South Australia demonstrated to Wilson the possible outcomes of land reform, but it was New Zealand that first suggested the practical contributions that exotic animals could make to this process. Wilson recorded in *Rambles at the Antipodes* that New Zealand had even fewer native animals than Australia, and that English grasses were displacing native pastures.[19] He admired the efforts that New Zealand colonists had made in successfully introducing pheasants. Wilson's observations while in New Zealand and his papers given before the PIV prompted him to think about the social, aesthetic, and scientific value of introduced organisms, but it was only after he arrived in England that he began to think in terms of acclimatization and to contribute to assembling a network of empire.

When Wilson arrived in England, he immediately began writing to the *Times* advocating for the redistribution of animals to Australia and elsewhere in the world. Wilson sought the aid of wealthy ex-colonists to transport alpacas and salmon to Tasmania, Victoria, and New South Wales. It was during this period that Wilson began to explicitly refer to the transportation of animals as acclimatization, drawing on the works of Isidore Geoffroy Saint-Hilaire and the Société Zoologique d'Acclimatation (SZA) and justifying why Australia, Britain, and the empire as a whole would benefit from acclimatization. In these letters to the *Times*, Wilson articulated for the first time an important principle of acclimatization within the Australian colonies and the empire, that the distribution of animals in the world was incomplete and through selective scientific acclimatization could be improved. Wilson believed that Australia possessed climatic conditions with the potential to support many different forms of life.[20] The empire's duty to harness this potential was a key to Wilson's enthusiasm for acclimatization. He believed that: "The Almighty has given us a wonderful variety of good things, and beautiful lands in which to place them; but the task of so distributing them seems to have been left to us, and that task we have sadly, nay sinfully neglected."[21] Other acclimatizers, such as George Francis in South Australia, also saw acclimatization as an integral part of enacting God's plan for Australia. Professor of natural science at the University of Melbourne Frederick McCoy and the naturalist George Bennett believed they were transferring "representative types" of species across different spheres of creation and therefore beneficially augmenting the divine plan.[22] Wilson, however, took this improving impulse and codified it, facilitating the import of exotic animals and export of Australian native animals.

The uneven distribution of animals in both Australia and Britain concerned Wilson, and he saw correcting this imbalance as part of the imperial civilizing process. This process was explicitly linked in Wilson's mind with scientific acclimatization when he quoted Isidore Geoffroy Saint-Hilaire, the president of the SZA, on how few animals have been domesticated, the importance of imported species to the development of agriculture and the potential agricultural value of thousands of wild animals that could be acclimatized.[23]

Wilson gauged support for acclimatizing animals in Australia by using the *Times* to solicit cooperation for acclimatizing a small flock of alpacas to Victoria and transporting salmon ova to Tasmania. He argued that as settlement expanded, alpacas were needed to graze areas that could not support sheep, thus explicitly linking acclimatization and colonization.[24] He believed the introduction of salmon to Tasmania was a useful test case for acclimatization and had the potential to facilitate a profitable commercial fishery. These experiments were noticed by naturalist Frank Buckland and the comparative anatomist Richard Owen and helped to catalyze the formation of the Society for the Acclimatisation of Animals, Birds, Fishes, Insects and Vegetables within the United Kingdom (SAUK).

Reacting to and inspired by Wilson's activities and the SZA, Richard Owen organized a special dinner on 21 January 1859 at London's Aldergate tavern. Key members of the English zoological establishment were invited to dine on roasted eland.[25] It is unknown whether Wilson attended, but he was certainly in London at the time and was known to be an acclimatization enthusiast, so he was probably invited. Eland are a species of large South African antelope that were first introduced into England by statesman and naturalist Lord Derby in the 1840s; he bequeathed his specimens to the London Zoological Society after his death.[26] The guests discussed the possibility of acclimatization in England, and concluded that eland was "the finest, closest, most tender and masticable of any meat."[27]

One of the attendees at the dinner was naturalist and Zoological Society of London secretary David William Mitchell. He was inspired to write a long article on acclimatization for the *Edinburgh Review*. The article began by detailing the history of the SZA and then discussed how the existing stock in the London Zoological Gardens could be used for acclimatization.[28] He continued by arguing for the introduction of deer from all over the world and detailing the successful establishment of a small herd of eland in England. The key virtues of antelope, according to Mitchell, were their potential as game animals, their beauty, and their "rapid growth, fecundity, and hardiness."[29] This early paper established three key characteristics of acclimatization in England and the empire: the value of exotic animals for field sports, ornamentation, and food. Mitchell's contribution to acclimatization was to be short-lived. Shortly after being appointed the director of the newly established acclimatization gardens in Boulogne, Mitchell fatally shot himself.[30]

Responsibility for acclimatization passed to more stable hands including John Crockford, proprietor of the *Field* newspaper, naturalists Frank

Frank Buckland. Frontispiece to George C. Bompas, *Life of Frank Buckland* (London: Smith, Elder, 1888).

Buckland and Professor Owen, sportsman Grantley Berkeley, and Wilson himself. These men were important players in the formation of SAUK. At this first meeting, it elected Buckland secretary, Berkeley vice president, and the Marquis of Breadalbane president. Wilson actively participated in the early meetings of the organization.

In August 1860, Wilson successfully moved a motion: "That the duty of forwarding this object [acclimatization] seems to rest particularly on Great Britain, possessing, as she does, dominions in every climate, and a mercantile marine of unrivaled extent and enterprise."[31] He went on to restate the potential of acclimatization in Australia, the relative poverty of animal species in the Australian colonies, and Isidore Geoffroy Saint-Hilaire's conviction that a fraction of the world's potentially useful animals were adequately exploited.[32] Wilson concluded that when "every good thing had been put into every good place, then they would know what a really magnificent empire they had."[33] These statements point out how SAUK was orientated toward the empire and the Australian colonies from its very inception.

In November 1860, Frank Buckland published "On the Acclimatisation of Animals" in the *Journal of the Society of the Arts*. Buckland was the son of William Buckland, the influential paleontologist and geologist. At this early stage of his career, assisted by his father and seeking to make a name for himself, he was serving as an assistant surgeon to the Second Life Guards. Later in life, he would become a popular science writer and inspector of the English salmon fisheries. His 1860 acclimatization article argued that Britain would benefit economically and agriculturally if climatically appropriate animals were acclimatized, and he argued passionately that wealthy patrons and the state should support SAUK.

Early Australian experiments in acclimatization, the culinary potential of Australian and African animals, and Wilson's organizational talents were central to Buckland's argument. Referring to naturalist John Gould's two books, *Birds of Australia* and *Mammals of Australia*, he argued that wonga pigeons, kookaburras, brush turkeys, and black swans should be acclimatized in England. According to Buckland and Gould, wonga pigeons were "first-rate birds of the table," and kookaburras (then referred to as laughing jackasses) were delightful creatures with an amusing song.[34]

BUCKLAND ALSO REFERRED TO Wilson's suggestion that wombats should be introduced to England as a "domestic animal not too large to be consumed by a middle-class family."[35] Australian animals, along with African antelope and Asian deer, were intended by Buckland to improve Britain's agriculture and hunting. Buckland acknowledged Wilson's influential work in raising "the public mind to the importance of acclimatization," experiments with Murray cod, and attempts to secure public funds for acclimatization.[36] He was particularly interested in Wilson's attempts at acclimatizing salmon in Tasmania and alpacas in Victoria. Buckland referred to a no-longer-surviving letter by Wilson concerning "utilising the waters" that argued that fish hatching "would pay," drawing attention to the commercial success of salmon fisheries in Scotland and Ireland.[37] Buckland also commented on Wilson's greatest early success in organizing acclimatization through bringing the cause to the attention of Lord Newcastle, the secretary of state for the colonies, who in turn instructed colonial governors to cooperate with British acclimatization societies.

By the end of 1860, a confluence of interests from colonists like Wilson, naturalists like Frank Buckland, and aristocratic hunting and fishing patrons had formed the metropolitan hub of an imperial acclimatization network. The SAUK was oriented toward the colonies, influenced by prominent citizens, but would soon become secondary to the colonial acclimatization societies.

One of the last things Wilson did before he boarded the ship back to Victoria was to spend £500 given into his trust by the PIV's Zoological Committee to buy useful animals for introduction to Victoria. The animals he chose to purchase are fascinating because in the absence of specific instructions, he had a free hand to buy those species he regarded as useful. Thus we are given an insight into Wilson's preferences, free from the influence of other acclimatizers. Wilson concentrated on purchasing wildfowl from all over the world because the swamps and lagoons in Victoria were full of wildfowl and had the potential to hold more; he expected wildfowl would come under increasing hunting pressure as the colony expanded.[38] He hoped that introducing new species would bring "variety to the sportsman and an additional item to the housewife," combining utility and recreation—a constant in Wilson's approach to acclimatization.[39] His selection also had a certain whimsical quality; he paid for fifty pairs of "pretty little brown squirrels" to be imported to Victoria out of his funds.[40]

Wilson returned to Australia in late February 1861 and immediately began coordinating acclimatization in Victoria. At the end of the month, he called the first meeting of the ASV. Exploiting the secretary of state's directive that colonial governors were to cooperate with formalized acclimatization societies, Wilson persuaded the Victorian governor, Henry Barkly, to chair the inaugural meeting. The ASV came into existence with a motion from Wilson: "That the extent of area, variety of climate, and comparative recency of settlement of Australia, invest the subject of acclimatisation with particular interest; and that the establishment of an Acclimatisation Society, on the basis of those of Paris and London, being highly desirable, such society be now established."[41] This motion encapsulated Wilson's views on acclimatization at this point, it should link together various colonies within the empire to their mutual benefit and that Australia, in particular, could be improved through acclimatization. The new society subsumed the previous Zoological Society of Victoria (ZSV), taking on its stock, finances, and most of its board of management. These board members and other new ASV members brought their own views and interests in hunting, science, and conservation to the ASV—subjects that held little interest to the practically minded Wilson. The confusion of influences and agendas within the ASV can be seen in its eight aims. They were lifted directly from the SZA, but the ASV soon made them its own.[42]

- For the introduction, acclimatisation, and domestication of all innoxious animals, birds, fishes, insects, and vegetables whether profitable or ornamental;
- the perfection, propagation, and hybridisation of races newly introduced or already domesticated;
- the spread of indigenous animals &c. from parts of the colonies where they are already known to other localities where they are not known;
- the procuration, whether by purchase, gift, or exchange, of animals, &c., from Britain, the British colonies, and foreign parts;
- the transmission of animals, &c. from the colony to England and foreign parts, in exchange for others sent through to the society;
- the holding of periodical meetings, the publication of reports and transactions, for the purpose of spreading knowledge of acclimatisation, and inquiry into the causes of success or failure;
- the interchange of reports &c., with kindred associations in other parts of the world, with the view, by correspondence and mutual good offices, of giving the widest possible scope to the project of acclimatisation;
- the conferring [of] rewards, honoury or intrinsically valuable, upon seafaring men, passengers from distant countries, and others who may render valuable services to the cause of acclimatisation.[43]

Previous discussions of acclimatization heavily emphasize the first aim because of the ecological damage caused by some of the ASV's importations, particularly trout and sparrows.[44] The third aim, relating to the spread of indigenous animals throughout the colony, and the fifth aim, regarding exporting Australian organisms, are of equal, if not possibly greater, importance than the first aim.[45] Looking at these aims in their totality, rather than focusing solely on organism introductions, reveals that the ASV was constantly struggling to reconcile the contradictions between them. Acclimatization in Victoria was more than just an importation program. It became over time an entire system of management encompassing native and introduced animals aimed at improving, restoring, and exploiting the local environment. Wilson developed this program in conjunction with the members of the executive of the ASV including medical doctor Thomas Black, Professor

McCoy, Ferdinand von Mueller, and Thomas Embling as well as sundry hunters, farmers, and fishermen and the animals themselves.

Thomas Embling, in 1856, convened a parliamentary inquiry into alpaca acclimatization, and Wilson experimented with alpaca acclimatization in 1858. State botanist and ASV vice president Ferdinand von Mueller was also an early alpaca enthusiast.[46] The ASV organized and expanded these early attempts at alpaca acclimatization.

Although somewhat bombastic, Wilson's hope of establishing alpaca farms in Victoria was neither unprecedented nor entirely implausible. Alpacas and llamas have been used in South America as sources of meat, for their fleece, and as beasts of burden for thousands of years. The Spanish conquistadores made several references to the cultivation of alpacas and llamas in their accounts of South America and were impressed by their hardiness and potential usefulness.[47] Phillip II of Spain kept several alpacas as a novelty. They remained underused for several centuries, with small numbers slowly filtering across to Europe.

Three interrelated factors in the nineteenth century catapulted alpacas from curiosity to potential fortune makers. First, new mechanized looms were developed in England that were capable of spinning alpaca fleece into cloth with "the appearance of silk," and a fashionable market in high-quality alpaca wool garments developed. From 1836 to 1843 some twelve million pounds of alpaca fleece were exported from Peru to England alone.[48] Second, Peru became a republic in 1820 and banned the export of live alpacas to protect its monopoly on alpaca fleece. Peru could not produce enough fleeces to fulfill demand, making establishing a new source of alpaca wool a potentially profitable venture. Third, Charles Darwin, in his voyage to South America aboard the *Beagle*, observed alpacas thriving in diverse and challenging terrain and enduring both frigid cold and desert heat.[49] This hardiness fired the imagination of British landowners who thought that alpacas could graze otherwise unprofitable British wastelands and feed and occupy the teeming masses. Small-scale experiments in alpaca breeding ensued, but they were frustrated by the inability to acquire large enough alpaca herds to establish stable breeding populations.

In May 1861, £2,000 was made available to the ASV to introduce alpacas into Victoria.[50] The ASV had two possible suppliers: James Duffield and Charles Ledger. Ledger was a Peru-based English trader and adventurer, who in 1858 smuggled the first alpacas out of Peru over the Andes to Chile before delivering the alpacas to Sydney in 1858.[51] Duffield offered to export alpacas from South America and sell them to the ASV for £100 each.[52] The ASV declined this offer as being too expensive. Duffield decided to proceed to South America anyway, confident he would find buyers for alpacas once he returned to Australia, as he considered Victoria the best potential alpaca country in the world.[53] Ledger was more cautious, wanting government assurances and preferably upfront payment before he would proceed to South America. Negotiations continued, delaying the delivery of alpacas until 1862.

Despite the difficulties of acquiring alpacas, Wilson remained an enthusiastic supporter of their acclimatization and of both Duffield and Ledger. He compared Ledger and Duffield to the explorers Burke and Wills because both parties had the potential to open up vast tracts of the country for agriculture: Burke and Wills through the discovery of new lands, and Ledger and Duffield by providing the animals for filling these new lands.[54] Referring to the importation of alpacas, Wilson believed that: "Millions upon millions of acres of land are now lying absolutely waste and useless, over which it is believed that this animal [alpacas] would profitably depasture. From vast tracts of country covered with coarser herbage, and badly supplied with water, the bush fires annually sweep off myriads of tons of a kind of sustenance held to be eminently suitable to the alpaca."[55] He hoped that the combination of exploration, land reform and the introduction of alpacas would allow wastelands to be appropriately developed and bushfires reduced, transforming and improving Australia. The alpaca was an animal with the potential to embody Wilson's ambitions for agriculture in Australia. Unfortunately, Wilson's ambitions were not realized, and despite initial success, by February 1864 all the alpacas purchased from Duffield and Ledger were dead.[56] Wilson resigned as president in September 1864 and returned to England to live out the remainder of his days. In October 1864 the ASV shipped the remaining alpacas (purchased from the remnants of the NSW government flock) to Gippsland, and would never again experiment in acclimatizing the species.

Wilson's separation from the day-to-day activities of the ASV started in March 1862 when he traveled to England (in his absence Dr. Thomas Black served as the ASV's acting president) to get his cataracts treated. Active participation in the ASV became impractical. The growing gulf between Wilson and the ASV also reflected philosophical differences concerning the purpose and scope of acclimatization. Wilson remained convinced that acclimatization was best used to encourage small-scale intensive agriculture, both practically and symbolically.

In England, Wilson corresponded in detail with the ASV's council. In May 1862, the ASV voted Wilson £500 to purchase quadrupeds and birds for export to Victoria.[57] The ASV provided Wilson with a list of organisms they wished him to acquire for acclimatization: red deer, roebuck, pheasant, partridge, grouse, rook, blackbird, thrush, starling, lark, sparrow, redbreast, goldfinch, and chaffinch.[58] This original list presented a balance between different purposes: creatures to be hunted, ornamental animals, and those species that might control pests. However, the list was soon altered in favor of game birds such as partridges and grouse, and objecting to the acclimatization of birds "being rather ornamental than useful."[59] In particular, the ASV council thought that starlings were not only useless but injurious. Wilson objected to the removal of starlings from the list because they were ornamental and useful birds.[60] However, the rift between the ASV and Wilson continued to grow, and the council was determined to control which animals it acclimatized. Putting this principle into action, the council decided to destroy the ASV's monkeys, and appointed a subcommittee consisting of McCoy, Mueller, and Alfred Selwyn (the Victorian government geologist) to audit the animals in the ASV's possession and "report what animals can be got rid of as useless."[61] Wilson objected to the monkeys' destruction because he thought they had aesthetic potential and could amuse colonists as they strolled through Victorian parks. Arguing for extensive acclimatization, Wilson described himself as a "thorough acclimatiser" as opposed to a selective acclimatizer interested only in charismatic megafauna.[62] By this he meant that he was dedicated to acclimatization as a creative and progressive force.

Despite his increasing distance from the ASV's council, Wilson continued to advocate for "thorough acclimatisation" in polite society and the London press. His determination to spread wealth and plenty from Australia to England was not mere rhetoric. In 1862 Wilson acted as the ASV's agent in England. He took delivery of magpies, kookaburras, and wombats sent from Victoria for delivery to the Zoological Society of London, SAUK, and the SZA.[63]

By December 1863, Wilson's sight had improved, and he returned to Victoria. Resuming active control of the ASV, he gave a speech revealing that his lobbying had successfully persuaded the Admiralty to transport animals destined for acclimatization aboard its ships.[64] He also claimed credit for organizing a series of questionnaires sent to the British colonies asking colonial governors to cooperate in compiling lists of native animals suitable for acclimatization elsewhere in the colonies. Wilson also established contact with zoological gardens all over Europe, helping to further cement the zoological network of empire essential to the ASV's success.

Wilson's successor as president of the society was the former premier and arch Tory, William Haines, who led the ASV from 27 September 1864 until 24 October 1865.[65] Haines stated that through importing game animals, the colony might become "occupied, not merely by a mere collection of pauper agriculturalists, but by country gentlemen."[66] Haines's disdain for small farmers and enthusiasm for importing game animals illustrates the distance that grew between the land reform movement and the ASV after Wilson ceased being president. In contrast, while Wilson also wished for a hierarchical and stable agricultural society, he believed that it should be founded upon small holdings and yeoman farmers, and should resist the reproduction of Britain's monopolistic "country gentlemen." Dr. Thomas Black succeeded Haines; he served as president from October 1865 until 1872.[67] During these eight years, Haines and then Black diverged from Wilson's vision for the ASV; the ASV became interested in game animals, fisheries, and regulating the use of native animals.

WILSON'S INTEREST IN ACCLIMATIZATION was enmeshed in his passion for land reform and in his beliefs in promoting both the effective exploitation of Australian natural resources and a conservative democracy. In this way, he was similar to the French SZA.[68] He played a crucial role in establishing a network of empire that centered in the colonies rather than metropolitan London. Pushing the focus of acclimatization in the empire from the center to the peripheries increases the importance of understanding the strange position of "dependent independence" occupied by Australian science, environmentalism, and wildlife management.[69] Wilson can be seen as the ASV's man in England, lobbying for support and sourcing British animals not out of sentimental attachment but to ensnare British interest in acclimatization. His actions in England aimed to be a proof of concept for a grand cosmopolitan project where a shift from unconscious to conscious ecological imperialism would transform the British Empire.

Wilson's long absences overseas limited his influence on the day-to-day running of the ASV allowing other visions to manifest. The centrality of Wilson to the foundation of acclimatization in Victoria but isolation from its practical developments suggests avenues of exploration for the following chapters. He avoided theorizing about but acknowledged the scientific underpinnings of acclimatization, focusing instead on moral, social, and imperial imperatives. Thus to understand Victorian acclimatization practices, chapter 2 will focus on how scientists based in the Australian colonies theorized acclimatization and how their theories shaped acclimatization practices. In fact, all the following chapters investigate what other people did with the network of empire that Wilson worked so hard to create but did not control. In the process we will enrich our understanding of acclimatization as a social and scientific practice that both reacted to and advocated for change.

CHAPTER TWO

Local Acclimatization Theories

To comprehend and understand the trajectory of and contradictions within acclimatization in Victoria, it is necessary to understand acclimatization as a science embedded in time and place. On a broad imperial scale, acclimatization should be situated among contemporary debates regarding biogeography, Darwinism, paleontology, and resource conservation. Victoria in the 1860s offered its own particular challenges: understanding the precolonization distribution of mammals, birds, fish, and reptiles; conceptualizing transformations that colonization caused to the "balance of nature" in Australia; and what, if anything, colonists should do about these changes. It was imperative to understand these issues and theorize acclimatization to efficiently spend the ASV's importation budget, ascertain which animals will survive in Victoria and how they will improve, and or repair the local balance of nature. All of this occurred amid the foundation of local scientific institutions and publications, environmental change, strong overlapping patronage networks, and radically divergent approaches to natural history. Previous scholarship has focused on diverging acclimatization approaches in Europe and has ascribed Australian thinkers a secondary and subservient role.[1]

Three prominent and one obscure Australian naturalists used and reconfigured European theory and their own observations of Victorian natural history to justify and understand acclimatization. They concluded that it was the solution to releasing Australia's potential and restoring the balance of nature. These naturalists were Ferdinand von Mueller, Professor Frederick McCoy, George Bennett, and Henry Ridgewood Madden.

Mueller was the ASV's vice president from 1861 to 1872. He was a native German speaker and received his PhD from Kiel University for a thesis on the flora of southern Schleswig.[2] Soon after he arrived in Victoria, he was appointed government botanist (1856–1892) and used the position to advocate for the creation of state forests to conserve the timber supply and protect the local climate.[3] Mueller maintained steadfast opposition to Darwinian evolution for his entire life.[4] This was a characteristic he shared with McCoy.

As well as being the first professor of natural science at the University of Melbourne, Frederick McCoy served as the director of the Museum of Natural and Applied Sciences. Building on these roles, he served as an influential member of the ASV's governing council from 1861 until 1873 and acted as its chief zoological expert.[5] During this period he created a unique acclimatization theory.[6] He intended his theory to both justify and rationalize acclimatization in Victoria and to act as a bulwark against the spread of evolutionary thought within the colony. McCoy's lifework and acclimatization theory were an attempt to reconcile his observations of Australian wildlife and paleontology with European theorists and to negotiate complex patron-client relationships with European scientific luminaries such as Adam Sedgwick and Richard Owen.[7]

Owen was George Bennett's primary patron. They met in the 1820s and corresponded extensively over many years.[8] Bennett was a Scottish-trained doctor who immigrated permanently to Sydney in 1836.[9] Over the course of a prolific lifetime, Bennett published many scientific papers on Australian wildlife.[10] He was the founding secretary of the Acclimatisation Society of New South Wales (ASNSW) and passionate advocate for the idea that protecting, understanding, and commercializing native animals should be an important part of acclimatization.[11] In this capacity, he published two lectures on the mechanism, history, and utility of acclimatization.[12] The ASV republished Bennett's lectures in their proceedings and maintained an extensive correspondence with him.[13] He was an active supporter of McCoy's, influenced the ASV's actions, and emphasized Owen's influence on acclimatization in the United Kingdom and the Australasian colonies.

Henry Ridgewood Madden rejected McCoy's acclimatization theory. He was a Scottish-trained medical doctor, a homeopath, and an ASV council member.[14] He justified acclimatization using Darwin's *Origin of Species* in a paper presented to the ASV's council and published in the *Yeoman and Acclimatiser* newspaper titled "Acclimatisation as a Means of Restoring the Balance of Nature."

The actions of all these men must be seen in the correct local scientific and institutional context. The 1850s saw the establishment of important scientific and educational institutions in Victoria as colonists sought to understand and control their new home and reaffirm links to Britain. The University of Melbourne was established in 1853, initially focusing on a classical arts education but within a decade adding engineering, medicine, and natural science to the curriculum.[15] In order to better comprehend local conditions, the colonial government appointed Mueller as government botanist; Alfred Selwyn was recruited from the Geological Survey of Great Britain to run a local geological survey; and to round out the group, William Blandowski was appointed government zoologist. From 1854 Mueller coordinated the newly established Museum of Natural History. Mueller, Selwyn, and Blandowski conducted explorations, published government reports, and presented papers at local scientific societies. The first societies were the Victorian Institute for the Advancement of Science and the Philosophical Society of Victoria (1854). They rapidly merged to form the Philosophical Institute of Victoria. It published papers, organized events, and exchanged journals with scientific organizations in England, Europe, and North America. The collective result of all these actions was a small, lively but isolated scientific community.

The scientific center of gravity remained in Europe. Most scientific knowledge about Australia was produced by experts in the United Kingdom and continental Europe such as Richard Owen and John Gould. The ASV and the colonial scientists under analysis in this chapter drew upon, contested, and expanded European and colonial knowledge to theorize acclimatization in Australia. Richard Owen was a comparative anatomist of almost unparalleled repute in Victorian London, who aimed to centralize and control imperial comparative anatomy and paleontology and built part of his reputation from studies of Australia and New Zealand.[16] He published many papers on extinct and living Australian mammals.[17] Like many before and after him, Owen sought to explain why Australia was populated mostly by marsupials and not by placental mammals (with the exception of some native rodents, bats, and the dingo). Taxidermist-turned-entrepreneurial-naturalist, John Gould and his illustrator wife, Elizabeth, in 1848 completed the serial publication of *Birds of Australia*, for many years the definitive ornithological guide to Australia. In it, they noted the prevalence of the parrot family in Australia, the presence of "primitive" mound-building birds, and the absence of large gallinaceous birds.[18] Australia was strange yet maddeningly similar. Animals were missing that should have been present, and those that were there were difficult to explain.

Added to this was the difficulty of explaining biogeographical change and continuity in Australia and across the globe. Changing distributions of animals were apparent in colonized societies.[19] Whether change was attributed to evolution, as per Darwin's recently published *Origin of Species*, or to divine providence, it had to be reckoned with. The problem was particularly acute for acclimatizers who saw themselves as engineers of change. They asked, was nature to be changed to correct earlier colonial environmental damage, or altered in the name of encouraging the progressive elimination of primitive fauna that was doomed anyway? Our four naturalists worked through these questions when delivering public performances that were

Local Acclimatization Theories

Portrait of a young Frederick McCoy, 1837–1857. Courtesy of the State Library Victoria.

one part theorizing acclimatization, one part call for support, and one part defense in the face of criticism.

In November 1862, on the occasion of the ASV's first annual meeting, Professor McCoy addressed 150 of the colonial great and good who gathered at the Mechanics Institute to celebrate its early successes and to show their support. His talk celebrated acclimatization, articulated his acclimatization theory, railed against evolutionary justifications for acclimatization, and made suggestions about what animals to acclimatize in Victoria. To make these points, McCoy created distribution maps and brought in taxidermied animals to act as visual aids. Intellectually his arguments drew from, reconfigured, and combined with his natural history observations, the ideas of Louis Agassiz, Edward Forbes, and Richard Owen.[20]

McCoy began his speech by arguing that Isidore Geoffroy Saint-Hilaire was wrong to assert that organisms evolved to adapt to new climates when acclimatized in a new environment.[21] Previous analysis of this speech emphasized the first part, where McCoy stated that acclimatization entailed "the bringing together in any one country the various useful or ornamental animals of other countries having the same or nearly the same climate and general conditions of the surface," and explains his attitudes toward acclimatization through vague references to natural theology.[22] However, within the lecture, McCoy pointed out a problem with a key premise of British functionalism and natural theology, that species are perfectly adapted to their environment—and stated, "the commonly received notion, that all animals in all points of their structure are completely adapted to the external circumstances of their native habitat, must by no means be supposed to imply that in those parts of the earth where the temperature is the same, that the animals are the same; the contrary being notoriously the fact."[23] To explain the diversity and distribution of animal forms and to justify acclimatization, McCoy instead proposed paying attention to the "law of representative forms or species, of animals,"

> which is not only of the highest interest to the philosophical zoologist, but has a direct influence on acclimatisation. It is found that those parts of the world having similar conditions of surface and climate, but separated by natural obstacles one from the other, are inhabited commonly not by the same animals, but by "representative species," so like them in size, shape, colour, and habits, that to the common eye they are frequently identical, while the zoologist can easily prove that they are really distinct specific creations, originally placed, probably as a single pair, in the middle of the district they inhabit.[24]

McCoy was here making claims about the distribution of animals across the world, or biogeography, as it became known. Biogeography and the similarities and differences between old and new world organisms, including analysis of Australian fauna, were heavily studied and contested by mid-nineteenth-century naturalists as they hoped it would reveal underlying laws of creation.[25] Of particular relevance to McCoy and acclimatization was the question of "representative organisms."

Representative organisms were morphologically and functionally similar creatures that lived in distinct isolated geographical areas. To nineteenth-century naturalists, their existence raised the question of how such similar creatures could exist in such geographically diverse areas. Charles Darwin and the botanist Joseph Dalton Hooker both studied representative organisms. Hooker came to no firm theoretical conclusions; Darwin, however, explained representative species as members of genera that once had an extensive distribution but subsequently had a limited geographical range.[26] Owen saw representative species as reflecting transcendental affinities in the Divine Mind, not material descent.[27] These affinities accounted for functional similarities between Australian marsupials and old-world placental mammals; for example, possums performed the function of squirrels. According to Louis Agassiz and strongly supported by McCoy, these representative species originated in separate, physically isolated, zones of creation.[28] Edward Forbes, expanding on ideas of climatic species distribution first articulated by German naturalist Alexander von Humboldt, argued that physically isolated "representative forms" are systemically distributed by latitude and altitude.

All of this is somewhat confusing and nebulous. McCoy attempted to deal with this problem during his lecture by creating a world map that illustrated physical isolation, zones of creation, and representative species.[29] Unfortunately, no copy of this map survives today. It is possible, however, to reconstruct how McCoy used the map to make his points. First, he pointed out zones of latitude and how climate varies predictably moving north and south of the equator. Second, McCoy illustrated how mountain ranges, such as the Himalayas, and major oceans could form barriers to animal migration to otherwise climatically suitable areas. Third, he placed the distribution pattern of several important species with multiple "representative species" on the map, giving each species a single point of origin in separate zones of creation.[30]

Taxidermied specimens made it possible for McCoy to make his theories concrete for his audience and to illustrate just how similar representative species could be. He instructed his audience to observe "on the table the sacred crocodile of Egypt and the rivers north of the line, and a specimen of its representative species, from the rivers of the like latitudes in South Africa."[31] McCoy wanted his listeners to note how closely the crocodilians resembled each other in both habit and anatomy. Similarly, he brought several preserved birds of prey to his lecture so his audience could see how closely the European barn-owl and "kestral-hawk [sic]" were represented by Australian predatory birds living at corresponding latitudes "of almost exactly the same size, shape, habits, markings."[32] Forbes's influence can be detected through McCoy's insistence on the importance of corresponding latitudes. Owen's influence is apparent in the insistence that similarities of both form and function categorize the relationship between Australian and European birds of prey.

McCoy directly articulated the relevance of "representative species" to acclimatization when he postulated "the acclimatiser may, as you would theoretically expect, bring with absolute certainty of success *all* the representative species of any group into *each* of the localities [original italics]."[33] This statement represents a systemic break from the static view of nature embedded within British natural theology, within which he was educated, where form was supposed to fit function perfectly, and there should be no gaps in the economy of nature. It also clarifies an important element of McCoy's acclimatization theory, that the presence of one representative species in a geographic area and the absence of an equivalent representative species in a second area is prognostic of a species' ability to survive when acclimatized. Geographic isolation produced multiple representative species that could thrive in any number of climatically appropriate areas, and the "agency of man" could move representative species around "with the certainty of their thriving."[34]

Acclimatization theory and practice were interwoven. McCoy's theorizing had a concrete purpose and built upon the ASV's early successes and failures. He wanted to explain the successes and increase the odds of further success. Australia's great pastoral industry had, according to McCoy, succeeded because there were no native ruminants in Australia. Extrapolating from this and his theory of representative species, he reasoned that Indian deer species and African antelope could thrive in Victoria's dry and hot areas and that English deer would live comfortably on the temperate coast and the high mountains. He congratulated the ASV's successes in acquiring small groups of Indian deer and thought that gaining sufficient quantities of these animals to establish breeding populations would increase like the first wild horses introduced into South America and "increase (without domestication) and become permanently established with us."[35] Looking at this and all of McCoy's theorizing, it is apparent that his vision of acclimatization was all about studying Australian and worldwide fauna to supercharge and steer the ecological imperialism that had facilitated the conquest of the new world. There was no discussion of correcting colonial environmental damage or fixing the balance of nature in McCoy's work. The ASV's involvement with and advocacy for neo-ecological imperialism came from elsewhere.

George Bennett believed in environmental transformation through acclimatization, but he also advocated the careful protection of native animals as part of acclimatization because they were useful and ornamental. These opinions were very apparent when in October 1861 he performed "Acclimatisation: Its Eminent Adaptation to Australia" at the Australian Library in

Local Acclimatization Theories 29

Dr. George Bennett F.Z.S., 1840–1849. Courtesy of the National Library of Australia.

Sydney, hoping to persuade his fellow colonists to found an acclimatization society, and in his later publications on acclimatization and Australian nature.[36] To appeal to colonists, Bennett reminded them how their current prosperity was the result of introduced animals without which they would be "a wandering half-starved race, subsisting, like the aborigines, upon the produce of the chase, roots and grubs, and clothed in opossum, squirrel and kangaroo skins; for our turkeys, geese, ducks, fowls, our horses, cattle, sheep, hogs, &c., are the results of acclimatisation, as also the importation of donkeys."[37] Further prosperity was linked explicitly to acclimatizing plants and animals from India, Africa, and South America. Bennett shared McCoy's emphasis on climatic suitability and enthusiasm for exotic ruminants. Over the course of his long career, he had visited and written about the fauna and flora of Sri Lanka, China, Singapore, Mauritius, Indonesia, Papua New Guinea, and New Zealand.[38] The knowledge gained on these trips was supplemented by an 1859 visit to the United Kingdom, where he reacquainted himself with Owen, attended lectures, and visited the Zoological Society of London's zoological garden.

Combining all his various experiences led Bennett to recommend that Australian acclimatizers should focus on several areas. European wild boar, South African river hogs, Indonesian babirusas, Malay wild ox, and the domestic ox of Bali should and could supplement the existing cattle and pigs already in Australia.[39] The Zoological Society of London's (ZSL) successful eland importation and breeding program inspired Bennett to think expansively about antelope acclimatization.[40] Due to these experiments and owing to similarities of climate between New South Wales and southern Africa, he was enthusiastic about the prospect of acclimatizing several antelope species in Australia. With regard to exotic birds, his own experiences in China, a lecture given by John Gould in England, and some successful English cross-breeding programs led him to believe that the Japanese Green Pheasant (*Phasianus versicolor*) and the Chinese Ring-necked Pheasant (*Phasianus colchicus torquatus*) would thrive in Australia.[41] More broadly, because Australia lacked large native gallinaceous birds and possessed diverse climates, Bennett believed that the grouse and partridges of India, Sri Lanka, and the Himalayas would thrive in different parts of Australia.

For acclimatization to be successful in Australia, Bennett argued for paying attention to native animals as articles of trade, beauty, and food. In his lecture "Acclimatisation: Its Eminent Adaptation to Australia," he waxed lyrical about native animals. Quoll and possum fur could be used to create "excellent and warm socks."[42] Wombat flesh was "a great treat," kangaroo was a little dry, and roasted bandicoot could be favorably compared to "suckling pig in flavour."[43] The legs of brush turkey made excellent eating, and their eggs were "delicious."[44] Dining on brush turkeys could be supplemented with wonga and bronzewing pigeons and any number of water birds. For all his gastronomic gusto, Bennett believed passionately in the aesthetic value of native birds, the "butterflies of the vertebrate animals."[45] He was particularly enamored with bowerbirds, Australian parrots, and kookaburras. Protecting these birds, among many others, was in the colonies and acclimatizers' self-interest. Bennett hoped "to impress upon the public in this colony the necessity of preserving birds to a certain extent, so as to fulfil what nature has ordained with infinite wisdom and care, the equalisation of the races, and of obtaining a knowledge of their habits and economy, which will be found valuable to man as regards his comfort, as well as affording him security from important depredations. Many, regardless of this, are continually destroying useful animals, and, become thereby the means of permitting those of a noxious kind to increase."[46] Humans, when acting in a rash or foolish manner, could disrupt the harmonious balance of nature. He attributed a recent plague of locusts in Sydney (1856) to soft-billed birds being "ruthlessly killed or driven away."[47] Additionally, it was clear to Bennett that emu populations were in decline and that the brush turkey and lyrebirds risked being "numbered with the extinct birds" like the dodo and so many others.

Bennett's *Gatherings of a Naturalist in Australasia* (1860) explored his fears about extinction and disrupting the balance of nature in greater detail. Ecological disruption was linked to colonial dispossession of Aboriginal people because "Kangaroos, Emeus [sic], and other wild animals, forming the principal food of the aborigines, being recklessly destroyed by the settlers, led to serious complaints and outrages on the part of the former, who considered it just recrimination to destroy their sheep and cattle; hence a series of fatal feuds was raised between both parties, which will eventually end in the extermination of the black race also."[48] Treating Aboriginal people as primitives inevitably doomed to extinction was a very common colonial motif.[49] Bennett used it as a warning to colonists: reform your wastrel and destructive ways, or you too will be as doomed as the Aboriginals. This warning was reinforced with two examples, both concerned with controlling snakes.

A "lady residing in the vicinity of Sydney" told him that she was afraid to let her children play outside because of all the snakes.[50] She did not connect the snakes' presence with her habit of destroying the small hawks that visited her garden. Taking Bennett's advice, no more hawks were shot, and the children were free to run around. On a similar note, Bennett observed that settlers once destroyed kookaburras (*Dacelo novaeguineae*) under the mistaken belief that they destroyed poultry. However, settlers "having learnt, both by experience and by the observation of the naturalist, the utility of this bird in the oeconomy of nature, it is now rarely or never molested."[51]

Optimism and learning were critical to Bennett's understanding of acclimatization. He believed that, with sufficient study and forethought, it should be possible to introduce climatically appropriate and useful exotic animals while also protecting native animals and the balance of nature. Ecological imperialism could be harnessed and made conscious, and neo-ecological imperialism could help ensure that the colony remained prosperous and vital. Bennett did not theorize about reconciling these two propositions, neither did he make any international comparisons.

Of all the scientists who sought to codify and guide Australian acclimatization, Ferdinand von Mueller envisioned the most dramatic transformations—rain in the deserts, new rivers, renovated forests supporting animals from the world over—and was the keenest to situate Australia within international developments. These two propositions are linked. Drawing on the influen-

Ferdinand von Mueller in 1864. Courtesy of the State Library of New South Wales.

tial American conservationist George Perkins Marsh and the naturalist Alexander von Humboldt, Mueller believed that animals and plants were distributed via altitude and that climate could be engineered through afforestation, creating new habitats and engineering the balance of nature for human benefits.[52] He combined these ideas with a passion for Australian botany and a concern that colonists were squandering their natural resources.[53] All of this can be explored by looking at a series of talks that Mueller gave from 1859 to 1872, beginning with two he gave at the Philosophical Institute of Victoria (PIV).

Mueller viewed the PIV and its headquarters as "the pillars of a temple, in which, we trust, Science will reign for centuries."[54] When the PIV constructed the first stage of this temple in 1859, Mueller lavished praise on a picture of the explorer James Cook that was affixed to the wall. It showed "his eagle eyes in inexpressible delight" staring out toward Australia and, in Mueller's view, envisioning a transformed continent full of cities and farms and presaging the PIV's mission.[55] With the aid of science, industry would thrive; isolation would be overcome; geographers would map the blank spaces, currently regarded with "mingled feelings of dread and hope";

numerous artesian wells would transform the vegetation of the desert; and "game in manifold variety shall roam through the forest, in which the feathered tribes of many zones shall, in their melodies, have added to its primeval charms."[56] Mueller, like McCoy, believed that the best furred and feathered additions to the Australian wilderness should come from Africa and South America.[57] These animals, he believed, would allow colonists to "by peaceful conquest" create "another Indian empire in continental Oceania."[58] Australia could, through the selective introduction of appropriate plants and animals from the world over, be transformed into a mercantile powerhouse whose exports could rival India.

Adversity just made Mueller more dedicated to acclimatization, more expansive in his transformative ambitions, and ever more determined to prevent the squandering of Australian natural resources. By 1868 Mueller's running of the botanical gardens was being criticized, and the ASV was coming under fire when some of its importations became agricultural pests. His two papers "Forest Culture in Its Relations to Industrial Pursuits" and "On the Application of Phytology to the Industrial Purposes of Life" can be seen as defenses of Humboldtian economic botany and acclimatization.[59] In both, he argued that deforestation harmed the local ecology and soil quality and reduced rainfall. Recent droughts and floods, Mueller thought, were a warning of future environmental change, and he called on the public to "remember why the absence or destruction of forests involves periodic floods and droughts."[60] He further postulated that deforestation damaged even those trees left standing after farmers destroyed a forest. They became vulnerable to attack by "coleopterous and other insects" and infestations of mistletoe, "there being no longer a multitude of native birds in populous localities to devour the mistle [sic] berries."[61] Thus Mueller recognized the damage that colonization had caused to the economy of nature.

Mueller did not despair of or reject industrial progress; instead, he was optimistic that ecological restoration and renovation were compatible with industrialization and ultimately beneficial to humanity. When discussing the possibility of introducing Norwegian spruce to Australia, "Nature" he wrote, was "ever active and laborious, ever wise and beneficent—allows the tree thus to live, thus to convert the solid boulders finally into soil, and all the time adds unceasingly to the treasures of the dominions of man."[62] He envisioned introduced species of trees flowering in the desert and increasing rainfall, transforming the climate of semiarid districts within Victoria. While presenting "Application of Phytology," Mueller asked his audience to translocate ourselves now for a moment to our desert tracts, changed, as they will likely be, many years hence, when the waters of the Murray River, in their unceasing flow from snowy sources, will be thrown over the back plains, and no longer run entirely into the ocean unutilized for husbandry. The lagoons may then be lined and the fertile depressions be studded, with the date palm; fig trees, like in Egypt planted by the hundreds and thousands, to increase and retain the rain, will then also have ameliorated here the climate; or the white mulberry tree will be extensively extant then instead of the Mallee scrub.[63]

This vision of the future is one of extensive, benign, and managed environmental transformation. Environmental change was, to Mueller, necessary and beneficial, and any adverse effects could be offset by proper management and the careful introduction of exotic organisms. To Mueller there was no conflict between ecological imperialism—colonization through ecological transformation—and neo-ecological imperialism—maintaining colonies and correcting environmental damage through moving resources from elsewhere; both could and should be pursued simultaneously.

Debates about protection, restoration, and improvement were occurring on both a local and an international scale in the 1860s. The Victorian colonial government prepared a report on forest protection in 1865.[64] Locally, both the *Age* and the *Argus* newspapers published concerns about the connection between forest clearing and rainfall. One particularly apt editorial described it thus:

In a new country such as this, one can see to what extent man is a levelling agent, and how easily he can disturb the harmonies of nature. Mining, ploughing, road making, the cutting of drains, the formation of tracks, all aid in diminishing the conservative powers of the natural herbage, and it is not surprising that our best streams, such as the Loddon, Campaspe, and Avoca, are fast becoming mere channels for the efflux of sludge and sand. Even in those parts not touched by the gold-miner, the rivers are rapidly changing their character. The mere occupation of the country for pastoral purposes has produced great changes, and it is well to consider whether anything can be done to compensate for, if we cannot check, this kind of devastation.[65]

This statement demonstrates an awareness of the environmental damage caused by the gold rush and a desire to rectify the damage to the "harmonies of nature" by introducing new species that extended beyond the ASV itself.⁶⁶ Previous authors have pointed out links between these local impulses to restore the environment through acclimatization, the ASV, Mueller, and a broader international ethic of environmental renovation.⁶⁷ Specifically, it has been stressed that both the American naturalist George Perkins Marsh and Mueller supported the hydrological theory of climate change. In essence, Marsh argued that wantonly destroying forests reduced rainfall and this adversely affected the local climate.

Marsh was not working in isolation. He was part of a broader cultural tradition that had noted the decline of wildlife and forests in America. The same year that Marsh was composing a report advocating fish culture to restore declining fisheries in Vermont, another report was being written in Maine advocating restoring nature through legislative control of fisheries.⁶⁸ As early as 1837 in Ohio, the naturalist Jared P. Kirtland was arguing for restoring the balance of nature by introducing elk, domesticated deer, and buffalo to replace those species being lost to extinction.⁶⁹ The result of the developing conservation ethic in the British Empire and across the English-speaking world was that the ASV could draw on a body of theory and evidence that acknowledged environmental damage and saw the introduction of new species as a possible corrective. This ethos was perhaps articulated best in Marsh's *Man and Nature*:

> The equation of animal and vegetable life is too complicated a problem for human intelligence to solve, and we can never know how wide a circle of disturbance we produce in the harmonies of nature when we throw the smallest pebble into the ocean of organic life.
>
> This much, however, we seem authorized to conclude: as often as we destroy the balance by deranging the original proportions between different orders of spontaneous life, the law of self-preservation requires us to restore the equilibrium, by either *directly returning the weight abstracted from one scale*, or removing the *corresponding quantity from the other* [author's italics].⁷⁰

Marsh's concept of rebalancing equilibrium through animal introductions is of critical value when attempting to understand the ASV. Furthermore, and on a much deeper level, renovating nature implies an understanding that nature is not purely a self-regulating providentially guided system, but requires humanity's intervening hand to function properly.

Mueller's love of acclimatization can be best understood as a fusion of Humboldt and Marsh's theories with McCoy and Bennett's enthusiasm for Asian and South African biota and his own belief in the transformative potential of applied science. Careful study and experimentation would, he hoped, create an environment that was productive far beyond its envisioned Australian, European, African, and South American components and would enable Australia to become the jewel in the crown of the British Empire. For all Mueller's love of theory and international exemplars, he never applied Darwinian theory to explain environmental change or justify acclimatization in Australia. Given his known hostility to evolution, this is perhaps unsurprising. Just across the Tasman in New Zealand, Darwinian evolution was the prime justification for the necessity of acclimatization.

DR. HENRY RIDGEWOOD MADDEN tried to make Mueller, McCoy, and the broader ASV membership see how useful Darwin's evolutionary writings were for understanding environmental change in Victoria and for thinking about what organisms the ASV should introduce into the colony. His big opportunity came in August 1864 when he was invited by the ASV's ruling council to present a paper at a public meeting held at the Mechanics Institute. It was part of a monthly series given by a series of experts throughout 1864. These included works on the Victorian fisheries and the virtues of Indian game birds. All of them were published as additions to the ASV's 1864 annual report, except for Madden's paper. It is possible to speculate that McCoy used his influence in the ASV to block the publication of Madden's paper because of its Darwinian leanings. It is also possible that the ASV did not want to offend its patron and key supporter, the ardently antievolution Victorian, then Mauritian, governor Sir Henry Barkly.[71] The ASV certainly did not exclude Madden from its activities because of his views. He served on its council for over a year, regularly attended meetings, was appointed to a three-man subcommittee to help organize its importation program, and when he left the colony was made an honorary member.[72]

Madden delivered his paper as ASV council member in good standing, committed colleague, and enthusiastic acclimatizer. It was titled "On Acclimatisation as a Means of Restoring the Balance of Nature" and began not with Darwin but by evoking the image of an enclosed aquarium full of plants, fish, and mollusks that held each other in "equipoise" demonstrating in miniature the "balance of nature."[73] Madden believed that the public was deeply indebted to Darwin for illustrating "the mutual dependence of various organic forms upon each other; the complex nature of this arrangement, and the readiness with which it is disturbed."[74] He pointed out the centrality of the struggle for existence in Darwinian thought and argued that left undisturbed, the struggle in any particular location "gradually settles down to a state of equilibrium." Colonization, urbanization, and the advent of farming disturbed this equilibrium. It created new foodstuffs for locusts, disturbing the range of native birds, and thus leading to "insects not in abundance, but in super abundance."[75] The paper was read during the last months of Edward Wilson's presidency; he attended the meeting and concurred that insect pests were abundant in Victoria as an unintended consequence of colonization.[76] When Madden finished speaking, Wilson asked for comments from the audience. Gideon Lang (Victorian pastoralist) responded by saying that kangaroos had multiplied out of control since farmers started killing dingoes and that Aboriginal people's demographic collapse led to decreased hunting pressure on possums and an increase in their numbers. He concluded by speculating that for "every European plant they [farmers] introduced they [the ASV] would have to introduce a predator" and fervently hoped the ASV would be up to the task.[77]

ALL THE PEOPLE DISCUSSED in this chapter believed that it would be necessary to introduce small insect-eating birds to Victoria. They also supported the introduction of hares to the colony. Despite differences over the mechanism of acclimatization and shifts in emphasis, all of the ASV's scientific leadership shared a set of common beliefs: Victoria is bereft of species and we must fix this problem to aid colonization; and colonization has broken Victoria, and we must fix it to protect the future of the colonial project. Sparrows and hares could address these problems.

Both private individuals and the ASV introduced hares to Victoria, and they became one of the ASV's early success stories. The involvement of the ASV in their introduction is, however, difficult to explain given that hares were known to be agricultural pests and the ASV was interested in intensive yeoman agriculture. William Lyall introduced hares to Victoria in 1858 or 1859, several years before the formation of the ASV. He was a squatter but had a keen interest in intensive and scientific farming. This led to studying agricultural chemistry, extensive experiments with irrigation on his properties, and breeding prize Hereford cattle.[78] Once the ASV formed, he became an active member, and participated in the Victorian Agricultural Society and the Victorian Racing Club. At first, the hares introduced by Lyall struggled. They soon, however, began to breed and thrive, a sure sign for the ASV that their acclimatization was possible. In mid-1862 the ASV imported three pairs of hares, courtesy of Dr. Philip Lutley Slater, secretary of the ZSL.[79] In July 1863, two pairs were sent to Phillip Island, where the ASV possessed half a square mile of land as a breeding depot. The Phillip Island site was chosen because its "freedom from native cats [now commonly known as quolls]" made it safe for young rabbits and hares.[80] The remainder of the hares stayed at Royal Park, where they began breeding. Some of the resulting progeny were given to James Henty for his Portland station.[81]

Unfortunately for current scholarly debate, the ASV has not left many written records indicating precisely why it decided to attempt the acclimatization of hares. We might, however, speculate that the virtues of hare acclimatization seemed self-evident at the time. It is possible to reconstruct some of the ASV's reasoning by looking at records relating to the acclimatization of species closely related to hares. In the *Answers Furnished* survey, the ASV wrote that it desired to introduce Cape hares, jackrabbits, chinchilla, and the spring hare of South Africa to increase the food supply in inhospitable areas.[82] It is not unreasonable to speculate that earlier ASV attempts at introducing hares were concerned with increasing food supplies to areas with milder climates suitable for hares.

McCoy certainly considered hares to be an appropriate species for Victoria's climate and regarded the successful establishment of rabbits and hares as proof of his acclimatization theory.[83] He also justified importing hare-like species by expanding on Owen's argument from "Report on the Extinct Mammals of Australia," contending that there was a transcendental relationship between hares and Australian hare kangaroos. Thus, he argued:

> The great grass-eating kangaroos have much of the habits, as well as the food, of the herbivore, which man is introducing in his place; the little *Hapaloti* have the size, shape, gait and general habits of the jerboa of Africa; and the little animals forming the genus *Logorcheates*, or the hare-kangaroos as they are called, are most singularly apt representatives amongst marsupials of the hares and rabbits in size, shape, colour, quality of the fur, living in the grassy ranges and plains, and sitting in forms like the common hare, which, when introduced by man (as well as the rabbit), thrives in the same places even better than the natives.[84]

McCoy published his expansion of Owen's arguments in the Australasian newspaper under the pseudonym Microzoon.[85] He also used his pseudonym to argue for the validity of Owen's argument from "On the Geographical Distribution of Extinct Mammalia" concerning the functional adaptation of

Local Acclimatization Theories 39

marsupials to drought.[86] McCoy's use of both the content and implications of Owen's publications illustrates the importance of Owen to McCoy's acclimatization theory. Maintaining that there was a transcendental relationship between hare kangaroos and hares also demonstrates that McCoy saw the successful acclimatization of rabbits and hares in Victoria as confirmation of his theory.[87]

The rest of the ASV also saw the early successful acclimatization of hares in Victoria as a proof of concept. By 1864 the ASV's annual general report stated that the hares were doing well.[88] In fact, by September 1865 it was remarked that the hare population was increasing "extraordinarily rapidly."[89] That year the ASV's annual report asserted that the "English hare may be regarded as fully established," and that the animals at Royal Park had "bred repeatedly."[90] At Phillip Island, the hares increased so rapidly that William McHaffie, the principal squatter on Phillip Island, estimated their number exceeded 200.[91] Soon after McHaffie's letter, hares were caught at Phillip Island and distributed to the rest of the colony at the price of £2 per head.[92] Hare acclimatization demonstrated the ease of zoological acclimatization and its potential to improve the food supply and encourage outdoor recreation.

Sparrows were another test case for the ASV. They were introduced into Victoria by the ASV in 1862 from China and from England.[93] They were one of five species of birds introduced because of the ASV's belief that they would control crop-destroying insects. McCoy wrote: "It may be mentioned generally that Victorian farmers and gardeners suffer very much from the depredations of insects, and therefore any of the soft-billed birds of Europe, or other temperate countries, are desired in unlimited numbers, particularly those, which like the robin (*Erythaca* [sic] *rubecula*), and hedge-sparrow (*Accentor modularis*),[94] love the neighbourhood of man."[95] He saw the acclimatization of birds as an opportunity to control agricultural pests and to fill empty representative types in the local "economy of nature" as indicated by transcendental similarities of both form and function. Madden, on the other hand, thought that sparrows should be introduced to Australia because colonization had disrupted the Darwinian struggle for existence.[96] McCoy was particularly keen on introducing secretary birds (*Sagittarius serpentarius*) from South Africa for "the destruction of our snakes, for which its structure is specially adapted."[97] He also lamented that Victorian forest trees were full of destructive insect larvae "while in the whole country there is no representative of the woodpeckers, appointed in other parts of the world to remedy this evil."[98]

When defending bird protection within "Acclimatisation: Its Eminent Adaptation to Australia," Bennett referred to a report to the French Senate, presented, in 1862, by Senator Bonjeau, on the importance of birds to agriculture.[99] The ASV later used this report to justify and defend importing sparrows to Victoria. A "board of practical men" in France wrote the report; they claimed that crops in France were poor because the peasantry destroyed insect-eating birds, and proposed that sparrows be protected in France.[100] The first section, the "Importance of Birds to Agriculture," argued for the protection of small insectivorous birds (i.e., sparrows), the destruction of birds that ate insectivorous birds (i.e., hawks and eagles), and the protection of birds that ate mice (i.e., owls).[101] Introducing sparrows was part of the ASV's mission to restore and improve nature. It fits well with the ASV's mission to study the local ecology and carefully judge significant absences and damage.

Sparrow introduction proved controversial from the very beginning. Farmers resisted it because they thought that the ASV was introducing an old country pest that they had hoped to leave behind. The ASV's friends at the Wilson-owned *Argus* leaped to the defense of the ASV, arguing it is "probable that, in all its importations, the Acclimatisation Society has not been guided by a perfect wisdom; yet the most ignorant of its detractors must surely know and understand the true object of acclimatisation. That object is not to disturb but to maintain the balance of life—to establish and not to violate the economy of nature. The introduction of civilized man into Victoria has rendered essential the introduction of other animals, to preserve that balance and to maintain that economy."[102] This statement was an almost verbatim regurgitation of Madden's acclimatization theory stripped of its more overtly Darwinian overtones. Ultimately it was the skeptical farmers who were correct. McCoy, Mueller, Bennett, and Madden failed to predict that sparrows would themselves become pests worse than the insects they were introduced to control.

All of the papers discussed in this chapter were performances enacted in the emerging scientific centers of colonial Victoria and New South Wales. They aimed to persuade colonists to support acclimatization and to demonstrate how it could improve their lives by filling gaps in, and emerging problems within, the balance of nature. They combined European theory and writings with their own observations of Australian nature. McCoy combined Agassiz, Forbes, and Owen to argue that there were vacant representative types in the Victorian economy of nature. Bennett drew on Gould, Owen, and the LZS to argue that Australian fauna was vulnerable, had value, and could be exchanged for useful animals from Africa, Asia, and South America. Mueller maintained a constant dialogue with the ideas of Marsh and

Humboldt to argue for radical environmental change and protection. Madden tried to make the others aware of the usefulness of Darwinian theory for explaining environmental changes in colonial Victoria. All of this is enmeshed in reactions to the Australian environment and is distinct both from how acclimatization was understood in France, England, and New Zealand and from how existing scholarship discusses Victorian acclimatization.[103]

THE ASV TOOK ITSELF SERIOUSLY as a scientific organization. The theories and programs proposed by McCoy, Bennett, Mueller, and Madden were shaped by and sought to influence environmental change in Victoria, the emerging scientific cultures of the Australian colonies, and contemporary ideas about environmental change and biogeography. They sought to understand why the distribution of animals in Australia was so different, if it could be improved, how colonization had damaged the balance of nature in Victoria, and what should be done about it. While they were divided by scientific theory, all four men believed, with some variation in emphasis, in an acclimatization practice that combined restoring and improving nature with animals from Europe, South America, Africa, and Asia. Acclimatization, as they envisioned it, involved simultaneous ecological imperialism and neo-ecological imperialism.

If acclimatization in Victoria was purely a science dreamed up by British, French, and Australian scientists to extend and repair the consequences of colonization, or a social ideal shared by Victorian and imperial reformers, chapters 1 and 2 would be sufficient to explain and possibly condemn acclimatization. Acclimatization was more than just scientific theory or ideological construction. Both the scientific practice and social support for acclimatization drew from beyond the ASV's leadership. The social and scientific practices of acclimatization was formed as much through interactions with the incipient acclimatization network as it was from abstract principles. Farmers, land reformers, fishermen, hunters, journalists, and politicians contributed to, transformed, and conditioned the social practice and science of acclimatization. Ecological change and the ASV's early activities shifted its purpose. Interactions and exchange of organisms between colonies generated new ideas and practices. Only by tracking the animals introduced by acclimatizers in Victoria, their interactions with other colonies, their successes and failures, and attitudes toward farming, fishing, and hunting can the trajectory of acclimatization in Victoria be truly understood and the relationship between theory and practice correctly interpreted.

CHAPTER THREE

Colonial Creations

In 1866 Sir Henry Barkly shared the grounds of Government House in Mauritius with ostriches from the Cape Colony and emus from Victoria.[1] He had previously been the governor of Victoria and was the Acclimatisation Society of Victoria's (ASV) patron. While serving as the governor of Mauritius, he continued to help the acclimatization movement by acting as a waypoint for African and Australian animals being shipped around the world. He also shipped "Mauritian animals" to Australia. Many of these animals were originally imported from India and Southeast Asia to Mauritius for hunting or pest control.[2] The transfer of animals between colonies, knowledge of colonial animals, and colonial environmental transformations helped shape the ASV's acclimatization program. Looking closely at the complexities of intercolonial exchanges of ideas, personnel, correspondence, and animals deepens our understanding of the subtleties of ecological imperialism.

The ASV spent several years and hundreds of pounds unsuccessfully attempting to import eland and other African antelopes to Australia and lamenting that the gourami (a pond fish species) it attempted to ship over from Mauritius kept dying midjourney in the equatorial heat.[3] Its members were ecstatic when the Indian myna bird successfully established itself in Victoria, hoping against hope that the birds would control the caterpillars that were destroying their crops. Also, the ASV exported Australian animals between British colonies and beyond. All this occurred in addition to the much analyzed and significant export of songbirds, trout, and deer from the United Kingdom to Victoria.[4]

IN 1863 AN EMPIRE-WIDE ACCLIMATIZATION SURVEY, authorized by the Colonial and Foreign Office and written by the duke of Newcastle, was sent out to colonial governors. Previously it has been assumed that it was organized by the Society for the Acclimatisation of Animals, Birds, Fishes, Insects and Vegetables within the United Kingdom (SAUK) and reflected metropolitan priorities.[5] Wilson, however, claimed that he was responsible for getting the duke of Newcastle to send out the survey.[6] This claim seems highly plausible given the questions the survey asked, the enthusiasm with which the ASV responded to the survey, and its attempts to acquire completed versions of the questionnaire from other colonies. The survey contained a series of questions about organisms suitable for export from each colony, and which organisms each colony would like to see acclimatized within its boundaries.[7] There was a strong preference for useful animals that could increase the food supply and contribute to the establishment of manufacturing industries and "whose constitution and habits offer a reasonable prospect of successful cultivation."[8]

The ASV delegated responsibility for completing the questionnaire to a subcommittee consisting of McCoy, Mueller, and Henry Madden.[9] Within the questionnaire, Mueller produced the sections on plants and McCoy was responsible for the zoological sections. Madden's voice is conspicuously absent. McCoy and Mueller dutifully compiled a list of Australian organisms for export and exotic animals they desired to import that are tabulated in the appendix (tables 1 and 2). The tables also list which animals from the questionnaire the ASV actually imported and exported. The ASV's list is very much a product of where acclimatization in Victoria stood in 1863–64. By this time the ASV had persuaded the Victorian public to support its endeavors and completed early test cases based on importing British fauna. Its officers were convinced of their theoretical and practical competence and were eager to exploit and expand exchanges between colonies. The ASV was at the apex of its influence, confidence, and ambition.

The species exported by the ASV were all valued by the prominent members of the ASV, significant natural historians like John Gould and Isidore Geoffroy Saint-Hilaire, and European acclimatizers such as Frank Buckland. The ASV attempted to protect many of these species through successive game acts (see chapter 5). The most common justification given for exporting Australian fauna was their food value, followed distantly by hunting and aesthetics. Despite the low priority given to aesthetics, black swans, primarily valued for their beauty, were the most exported species of Australian fauna, followed by kangaroos, emus, magpies, kookaburras, and Cape Barren geese.

It is worth exploring the three most popular Australian animal exports in some detail to explicate precisely why they were in demand. *Answers Furnished* described the black swan as deserving "attention on account of its ornamental appearance, but also for its down and for its flesh, which, when obtained from cygnets, is excellent food."[10] Bennett and Buckland both mentioned the fact that black swans had been successfully exported to England in the 1850s, where they bred profusely.[11] The primary appeal of black swans was as a symbol of the exotic, contrasting the black swans of Australia with the white swans of Europe.[12]

Kangaroos were a symbol of exotic, alien Australian fauna; acclimatizers exploited this fact while also promoting their practical potential. In *Answers Furnished* the ASV described kangaroo flesh as "scarcely equal to that of most other game, though their skin furnishes a good kind of leather. From the peculiarity of their form, and their eccentric movements, they would constitute a very interesting feature in parks; and from their speed they might furnish a valuable addition to objects of sport."[13] In contrast, the Société Zoologique d'Acclimatation (SZA) took a utilitarian perspective, hoping to establish kangaroo farming in France's more arid colonies. To this end, it offered a £30 reward to anyone who could maintain a group of at least six kangaroos and prove that they had been bred for at least two generations—spurring interest in kangaroo acclimatization.[14] The same reward was offered for the domestication of emus or rheas. The ASV itself believed that emus stood "foremost as a bird desirable for naturalisation in other similar climates. The great ease with which its transit can be effected [sic] when small, the fair food which the flesh of the young bird affords, the adaptation of this bird as well to a sub-alpine as tropical clime, its contentedness with very indifferent food, its great size, its abundant oil (used by the colonists for medicinal purposes), its harmlessness, the value of its eggs, tend all to recommend it for introduction into many other countries."[15]

The popularity of some of the Australian birds must be treated with some skepticism because exchanges of exotic animals were often made for unspecified Australian "native birds," and magpies and kookaburras were merely the most common and easiest birds to catch in the Melbourne area. The most common destinations for Australian animals were, in descending order: France, England, India, Mauritius, Russia, and New Zealand. Active acclimatization societies in France, Russia, and New Zealand; the presence of Sir Henry Barkly as governor in Mauritius; and the importance of the Zoological Society of London as a repository of zoological knowledge and collections explain this pattern. It is further evidence of the transnational, transimperial, and transcolonial nature of acclimatization. Acclimatization was not a one-way process of ecological imperialism; it was a far more complex process where native animals were valuable and in demand both locally and throughout the world because of their beauty and utility. On a more prosaic level, exporting native fauna also established and reinforced networks of exchange that facilitated the importation of exotic animals. Additionally, Victorian acclimatizers were not alienated from their surroundings; their knowledge of local and imperial zoology and climate led them to believe acclimatizing exotic animals would be viable.

STUDYING THE ANIMALS LISTED AS desirable within *Answers Furnished* (see appendix, table 2) is a powerful way of looking at the ASV's acclimatization program because it is a list of intent and imagination, not a list of results. Hindsight and long-term negative environmental consequences have clouded many studies of acclimatization in Australia. Scholars saw that British animals such as hares, rabbits, and red deer became established in Australia, and they assumed that the original intent was to establish British animals. However, the successful establishment of British animals in Victoria was a result of the ease of acquiring relatively large populations of British animals (thus increasing the founding population) and not a cultural preference for the notional nationality of the animals. Once rid of this supposition, the ASV's exotic animal acclimatization program looks very different. Examining the different species the ASV proposed to import reveals that it desired fifteen African species, seven apiece from England and India, five from Europe, four from South America, three from North America, two from Asia, and one from New Guinea.

By 1863 the ASV was attempting to import considerably more species from outside Britain than British species, a fact that complicates the idea that the acclimatization of terrestrial vertebrates in Victoria was a simple process of Britainization. The ASV desired British species as part of the fauna of the empire and because they might be conveniently acquired. There was initial interest in British species to capture the public's imagination, but when these early introductions were successful, the ASV's imagination, and ambition, expanded. The large number of African species desired demands explanation. Perhaps the only reason for acclimatization not being described by scholars as an attempt to Africanize Australia is the difficulty the ASV had in acquiring and shipping viable populations of African animals to Australia.

Twenty-two of the species that the ASV listed as desirable imports within *Answers Furnished* were mammals. Thirteen were birds, supplemented by eight fish species and two insect species. The preponderance of mammal species reflected dissatisfaction with the distribution of Australian mammals, vacant "representative species," and a preference for mammalian foodstuffs. The smaller number of desired exotic birds reflects higher satisfaction with the distribution of native birds, with the exception of the need to replace declining game bird populations and supplement insect-eating species.

Attempting to aggregate why the ASV wanted to import particular species reveals the overdetermined nature of acclimatization and the need for specific case studies. Nevertheless, certain overall trends do emerge. The desire to increase the food supply in the bush was mentioned with reference to eleven different organisms, the climatic suitability of exotic organism was argued for nine times, vermin and weed control were mentioned seven times, providing raw products for manufacturing was mentioned twice, and sport hunting was used to justify acclimatization only in the case of acclimatizing gazelles. There is a continuing tension between importing animals to fill gaps in Australia's faunal distribution to aid colonization and importing animals to correct damage caused by colonization. Many species were desired because of their potential as food combined with their assumed climatic suitability. This combination was seen as a direct aid to colonization. Thus, the ASV advocated, "The Spring Haas (Helamys Capensis) or Leaping Hare of the Cape, for sandy and stony desert tracts in the northern districts, is desired in the hope of adding to the very scanty food to be found by the explorer or pioneer in such localities, to which the habits of the animal are well suited; its flesh would prove a welcome meal to many persons engaged in pushing the settlement of this new country."[16] *Answers Furnished* recommended many species for acclimatization without ever explicitly explaining why they were desirable. To understand why the ASV spent so much time and effort on attempting to acclimatize specific species and to see how it made these decisions, it is necessary to examine the relationship the ASV had with particular colonies and their inhabitants.

THE PLAINS OF AFRICA, full of exotic and potentially useful animals, were part of the acclimatizing imagination from the very start. Frank Buckland in England enthused about the virtues of eland and springbok and used their successful shipment to England to demonstrate the economic and gastronomic potential of acclimatization.[17] George Bennett, Frederick McCoy, and Ferdinand Mueller were early enthusiasts for African animals. The SZA in France was keen on the acclimatization of ostriches from its inception. The ASV had its African moment in 1862. By this time its infrastructure in Royal Park was well established, early experiments had succeeded, and a corpus of acclimatization expertise and theory was being formulated.

The Cape Colony was an obvious and useful source of African animals and ideas for the ASV. It shared vital institutions, traditions, and personnel with Victoria but also points of divergence that provoked useful fission. Like

Victoria, it was a British colony that was heavily dependent on pastoralism and had a strong ethic of improvement, an emerging scientific culture, concerns about environmental degradation, and a varied climate that included temperate to arid regions.[18] Unlike Victoria, the Cape Colony had to deal with previous rounds of colonization by the Dutch and the continuing presence of Boers and the Xhosa and Khoisan people they had conquered, imitated, and mined for environmental knowledge. The animals of Victoria and the Cape were profoundly different, one devoid of ruminants and predators and replete with marsupials, the other full of predators and ruminant species whose numbers were in decline. Cape animals, their destruction, domestication, protection, and also their transfer to Victoria were understood through a combination of British, Boer, and African scientific and hunting traditions. The exchange of animals and knowledge between Victoria and the Cape was facilitated by two men, Sir George Grey and Edgar Leopold Layard, who had extensive contacts in and knowledge of the Cape Colony.

Sir George Grey was the quintessential imperial careerist and a huge acclimatization enthusiast. During his long and varied career, he was an explorer in Western Australia, the governor of South Australia, the governor of the Cape Colony, the governor of New Zealand (twice), an elected New Zealand politician, and a delegate to the Australian Federation congresses.[19] It is his time in South Africa and New Zealand that most materially affected the development of acclimatization in Australia and New Zealand. He turned an island off the coast of New Zealand into an acclimatization laboratory full of plants and animals from the world over.[20] He corresponded with the ASV and exchanged Indian pheasants and New Zealand birds for black swans, Cape Barren geese, and fallow deer.[21] Perhaps Grey's most significant contribution to the ASV was dragging Edgar Layard behind him in his wake.

The ASV first was first contacted by Edgar Layard in 1862 when he wrote to it, in his capacity as Grey's private secretary, asking if it could ship surplus fallow deer to New Zealand.[22] By this time his uncle Charles Layard was in contact with the ASV and was supplying it with animals from Ceylon. The Layard family members were officials in colonial Ceylon for several generations. Edgar's father, Henry, worked in the Ceylon civil service, and Edgar's uncle Charles was a district judge.[23] Edgar himself worked as a law clerk, then a barrister, positions that were not very taxing and that allowed him plenty of time to pursue ornithology and publish on the avifauna of Ceylon.[24] He then began his service as Grey's private secretary, first in the Cape Colony, then in New Zealand.[25] While he was in the Cape Colony, he was also the curator of the South African Museum, which allowed him to become familiar with the flora and fauna of the area. After leaving Grey's service, Edgar returned to South Africa via Australia.[26] During this journey, he attended an ASV meeting, where he made recommendations about the acclimatization of African animals and agreed to act as an agent for the ASV in Africa.[27]

Unfortunately, no records have survived concerning what animals Edgar Layard recommended in his 1862 meeting with the ASV. It is possible to reconstruct what they may have been because in 1864 the colonial secretary of the Cape Colony tasked Layard with completing the acclimatization survey.[28] Layard's response to the acclimatization survey is interesting because he had met with the leading members of the ASV and was aware of their aims and theories. Through this contact and his own experience, he was aware of the climate in Victoria and in New Zealand and the extent of environmental change caused by colonization. His list combined this insider knowledge of antipodean acclimatization with his observations and the distilled wisdom, traditions, and prejudices of several centuries of African, Boer, and English hunting and natural history knowledge acquired in the Cape.[29] It answered all the questions in the survey about the existence of animals in the Cape along with his comments on the practicality of acquiring particular species and their climatic suitability for England, New Zealand, and the Australian colonies. I have tabulated his animal list in table 3 of the Appendix, along with his comments about climatic suitability, use, and availability.

Layard's list is a chimera. It draws on multiple British, Boer, African, and Australasian hunting, farming, and scientific traditions and filters them through Layard's own experiences and understanding of the aims of the acclimatization movement. The first thing to note is the naming conventions that Layard uses. The common names that he uses are frequently of Boer origin, that is, hartebeest, bontebok. This practice was fairly standard and reflects the continuation and incorporation of Boer traditions in the British Cape Colony.[30] His scientific naming practices were deeply indebted to the publications of the naturalist and museum curator Andrew Smith and the game hunter William Cornwallis Harris.[31]

It is interesting to note which Cape animals did not feature within Layard's list. There are no mentions of any of the mammalian carnivores (lions, leopards, hyenas, jackals, African painted dogs) anywhere on the list. Partially, this is an artifact of the fact that at no point were any questions asked in the survey explicitly about carnivores. It is also, however, a product of the fact that both the Cape Colony and Victoria were deeply dependent on the pastoral economy that had long viewed predators as vermin to be exterminated.[32]

Colonial Creations 49

There are also patterns among those animals that Layard said existed in the Cape but were unsuitable for introduction in Victoria. He thought that elephants and hippos would not thrive due to a lack of sufficient lakes and rivers and an inability to digest eucalyptus leaves.[33] The other main reason he gave for not recommending importing those species to Australia was their aggression and inability to be domesticated. Conversely, the mammals that Layard did recommend for introduction were predominantly grazing animals that came from Cape environments that had analogs in Australia, and that could possibly be domesticated.[34] The emphasis on domestication draws on long-standing Cape traditions of desiring to privatize, control, and domesticate wild game.[35] Importing species to Australia that were already semi-domesticated was one way of reducing costs, streamlining the process, and overcoming the fact that many species were in decline in the Cape. Birds could be easily captured, domesticated, and sent to Australia in considerable numbers, making them ideal candidates for acclimatization. They were sought after because they would provide food and sport (pheasants and partridges), correct environmental damage (Cape sparrows), or control venomous native snakes (secretary birds). A few years later the ASV managed to introduce ostriches from the Cape Colony and France.[36] Cape experiments in ostrich farming inspired these attempts. Although ostriches could be induced to breed in Australian conditions, ostrich farming was abandoned after it was found to be economically unfeasible to compete with African ostrich farms.[37] Layard did not directly inspire these experiments, but his practical influence can be seen in how they were conducted.

Layard ended his response to the acclimatization survey with a series of practical suggestions for facilitating successful acclimatization. He emphasized the need to create acclimatization depots in the countries of origin and that these depots needed to get animals used to confinement and artificial feeding. Animals should then be transported in specially designed cages and accompanied by trained staff responsible for their care during the voyage. To facilitate the capture of larger animals in breeding pairs useful for acclimatization, he recommended that acclimatization societies combine to outfit an expedition because "the larger animals of South Africa are being exterminated or driven away into remote and almost inaccessible regions."[38] Despite this warning, Layard was optimistic about what animals he could supply to the ASV. The ASV's attempts to acquire African animals thus became a combination of its desires, Layard's inclinations, and the practicalities of acquiring live specimens in a colony with primitive infrastructure, extensive hinterlands, and rapidly declining faunal distributions.

The practical results of Layard's assistance and the ASV's African dreaming were minimal. He was able to organize multiple shipments of African pheasants and partridges, some of which survived the trip to Australia, but the ASV never liberated any.[39] Attempts to ship eland and zebra to Australia were vexed. On multiple occasions zebras were shipped to Australia from the Cape and France; unfortunately, it proved impossible to get a breeding pair to Australia, and the one male zebra they did acquire eventually died of old age in Royal Park.[40] Acquiring the various antelope species was a more complicated matter. In 1862 the ASV made £200 available for the acquisition of African antelopes. The Cape government placed advertisements in the *Government Gazette* that the ASV would be willing to pay for living specimens of eland, kudu, gnu, hartebeest, quagga, gemsbok, bontebok, springbok, and blesbok but that it should not be too hopeful because of the "difficulty in procuring proper specimens for acclimatization."[41] A simultaneous, better-funded, and better-equipped mission, organized by the Zoological Society of London, exacerbated this difficulty.[42] This mission consumed resources, undermined intercolonial cooperation, and ultimately diverted the few specimens that were captured to Great Britain. The ASV's subsequent attempts to acquire African antelope failed, with only the occasional eland specimen acquired, which often died in transport.[43] It was never possible to get enough antelope to breed them in the Royal Park facility, let alone sufficient numbers to allow release into the wild. The practical difficulties of acquiring and transporting sufficient quantities of animals meant that no African animals were ever successfully established in the wild by the ASV. The dynamics of this failed program reveals much about the ASV's climatic, cultural, and scientific imagination. Analyzing it helps draw out how Australian acclimatizers understand the distribution of animals across the world and how colonialism has changed colonies. Focusing only on successful species exchange would create an impoverished understanding of the ASV and its activities. Of course, analyzing successful intercolonial exchange programs, such as between Victoria and India and Ceylon, is also essential.

INDIA AND CEYLON WERE viable sources of animals for Australian acclimatizers for several reasons. First, there was the question of logistics. Animals could be shipped from Calcutta to Australia because of steamer lines that operated between the two colonies by the Pacific and Orient shipping line and broader trade in grain and horses.[44] Second, British imperial control of India, Ceylon, and Australia and active government support for acclimatization facilitated the transfer of animals. Third, Victorian acclimatizers could take advantage of colonial zoological gardens, museums, and private menageries as both suppliers of animals and customers for exported Australian fauna.[45] All of these institutions did not just passively make acclimatization possible; they actively contributed to acclimatization knowledge embedded within a contingent and fragile imperial network centered in the colonies. Knowledge and opportunity were created and intertwined by colonists who moved between Australia, India, and Ceylon; scientists with interests in these colonies; and books and articles published about the flora and fauna and climate.

Many individuals contributed to the movement of animals and knowledge to and from Australia, India, and Ceylon. Henry Edward Watts and George James Landells had lived in India and Australia and made suggestions to the ASV; these suggestions were acted upon and sometimes refuted by Indian scientists and collectors Edward Blyth, Edward Arthur Butler, Charles Layard, and Rajendra Mullick. Henry Watts was born and raised in Calcutta, where he worked as a journalist before immigrating to Australia in 1857 to try his luck at gold mining.[46] He became associated with the ASV through his work as the editor of Edward Wilson's newspaper the *Argus*. George Landells was second in command of the unsuccessful Burke and Wills expedition into central Australia.[47] He supplied it with camels and later sourced Indian animals for the ASV. Edward Butler was a soldier and an ornithologist; he published several books on Indian ornithology and engaged in the commercial animal trade.[48] Charles Layard was a prominent colonial British official and the uncle of Edgar Layard.[49] Rajendra Mullick was a prosperous merchant in Calcutta who maintained a private menagerie and helped keep the ZSL stocked with Indian fauna.[50]

Although all these individuals had interests in Indian animals, it took the intervention of Edward Wilson and the ASV to catalyze the exchange of organisms. Wilson returned from England to Victoria in 1861 to establish the ASV. During this voyage, he stopped at Colombo and wrote a letter talking about early acclimatization experiments in Victoria and asking that "every British community with any pretensions to intelligence or progressiveness" to form their own acclimatization society and aid others within the empire. As encouragement, he published a somewhat exaggerated list of animals that had already been introduced to Victoria or were currently in transit.[51] In response to this article, Blyth created a list of species he could supply. I have tabulated this list and Blyth's comments in the appendix (table 4).

Also, Blyth offered unspecified Indian carnivores and reptiles for the ASV's (nonexistent) zoological gardens in exchange for live or preserved specimens of Australian marsupials, monotremes, and parrots. This list is a good reminder that the ASV was not trying to build a network of exchange from scratch but hoping to take advantage of and adapt the knowledge and logistical support of existing zoological and museum networks. Blyth was correct in asserting that he "as curator of the Museum of the Asiatic Society" was afforded "peculiar facility in selecting Indian animals" for Australia.[52] He built this facility from the web of Indian collectors and collaborators he had assembled and the utility of the museum's collections for synthesizing knowledge about Indian species, their climatic distribution and diet, and for estimating which species would thrive in Australia.[53] Many of the species that Blyth suggested had until this point not been contemplated by any acclimatization society, and he opened the eyes of Australian acclimatizers to India's potential.

The ASV did not passively accept Blyth's recommendations, knowledge, or the animals he offered. Its leadership engaged in robust discussions with Blyth and other Indian experts based on their understanding of the Australian and Indian climates and of animals required in Victoria and their own slowly emerging knowledge of which Indian species thrived in Australia. The ASV declined most of the species that Blyth offered to send out and asked him to focus on supplying more axis deer and hog deer to increase their small herds.[54] It also made use of Blyth's connections to secure a small shipment of castor oil silkworms to conduct experiments in sericulture. Ultimately Blyth's contribution to the ASV's success was more about awareness and knowledge than actual animal importations.

By the time Blyth left India in 1863, the ASV was becoming more assertive about what it wanted from India and why. This assertiveness can be seen in the aforementioned 1863 Acclimatization Survey. Also, Henry Watts, in his paper "The Game Birds of India," argued that India was "the great market for animals in the Eastern world" and because of analogous soils and climate Indian animals may "fairly be presumed to be adapted to become denizens also of our continent."[55] Based on these similarities, and with Professor McCoy's support, Watts recommended that the ASV focus on importing red-legged partridges and black partridges.[56] McCoy and Mueller furthermore suggested focusing on Indian deer species both because of theoretical considerations and because the axis deer at the ASV's experimental farm at Royal Park thrived and even began to breed.[57] Thus both theory and practical experience were leading the ASV to create acclimatization knowledge as part of contingent and ever-renegotiated relationships between British colonies.

Blyth was not the ASV's only supplier. It sourced partridges and deer from Arthur Butler and Rajendra Mullick in India and Charles Layard in Ceylon. In March 1861 in its first letters to Charles Layard, the ASV reminded him that an acclimatization society was not a zoo, and accordingly they desired only "such animals as shall when set free be of service to the country."[58] And from Ceylon they desired peacocks, to kill snakes, and sambar and axis deer because they were hardy enough to survive the journey to Australia and to thrive once they arrived.[59] In response, Layard sent the ASV multiple axis and sambar deer, peacocks, two porcupines, one mongoose, and one "jungle cat."[60] This cat and the porcupines the ASV did not "much care for," but it continued to gratefully receive sambar and axis deer and partridges from Layard for several years.[61]

The ASV's relationship with Edward Butler was both problematic and revealing. In its initial letter, the ASV informed him that it desired only animals that would be "really useful" when turned loose and supplied him an assorted group of native Australian animals to initiate their scientific exchange.[62] Given these explicit instructions, the ASV was very surprised when Butler proposed shipping Bengal tigers to Australia.[63] In frustration, the ASV reminded Butler that there was no market for zoological curiosities in Australia and listed the species that the ASV did want from India in 1863: "Sambar, Axis and Hog deer, Indian and Chinese pheasants, Red Legged and Black partridges, Mino [sic] birds and sparrows [All of these animals appear either in Blyth's list, *Answers Furnished* or Watt's game bird paper]."[64] Butler continued to frustrate the ASV. Sometimes he supplied useful animals like hog deer, but at other times he sent out all-male groups of axis deer, a single antelope, "utterly useless" porcupines, and mouse deer.[65] These small deer were particularly irksome to the ASV because its experience had taught it that mouse deer were "too delicate for this climate."[66] The salient points to take from this are how rapidly assertive the ASV became in its knowledge and the difficulties acclimatizers had using preexisting zoological and commercial animal networks. It is also important to note how acclimatization evolved as a practice and series of ideas through the interactions of actors in India, Ceylon, and Australia. Acclimatization was a moving target, not a preconceived plan of action that was simply rolled out.

Unlike the failed Cape Colony project, the acclimatization of Indian animals was somewhat successful. Hog, axis, and sambar deer were successfully transported to Victoria, bred at the Royal Park depot, and released into the wild, where they thrived.[67] They can still be found in Victoria today.[68] Establishing Indian deer in Victoria was successful for a number of interrelated reasons: multiple suppliers, relatively large foundation populations, successful captive breeding and subsequent release in suitable areas, adaptable dietary preferences among the deer species, and somewhat successful legal protection. Indian myna birds were imported because they had developed a reputation as insect-destroying birds.[69] They are now a significant pest species.[70] By way of contrast, Indian pheasants and partridges never established themselves in Victoria despite multiple attempts over decades. The ASV's pheasant acclimatization program contrasts interestingly with an earlier ZSL attempt to acclimatize Indian pheasants in 1857. This attempt involved using multiple agents to gather hundreds of pheasants and put them aboard a ship for a very long journey to the United Kingdom.[71] The ASV gathered only a dozen pheasants and shipped them the relatively short distance to Australia.[72] Proportionally more pheasants survived the trip to Australia than Britain. Pheasants survived in Britain and became commercially available for gamekeepers who kept them alive as semidomesticated animals on game reserves. The Victorian pheasants bred in the Royal Park facility, but apart from some initial success of Phillip Island, were never able to thrive when turned out into the wild.[73] This vision of Indian and African animals surviving in the Australian wild and making the Australian wilderness both more understandable and survivable for colonists was crucial to the ASV's program.

THE PROGRAM WAS COMPLEX and evolved rapidly as Victorian acclimatizers learned from early experiments and took advantage of existing bodies of zoological knowledge in Britain, France, the Cape Colony, and India. It both was formed by and shaped intercolonial exchanges of animals, knowledge, and people. Without cooperation from the imperial government, the acclimatization survey would never have been sent out. Without the cooperation of colonial officials in the Australian colonies, India, Ceylon, and the Cape Colony, no animals would have been exchanged and *Answers Furnished* would have been blank.

Exploring the native species recommended for export within *Answers Furnished* and the species that the ASV did, in fact, export revealed extensive demand for some native Australian species, including black swan, emus, and kangaroos. It also further demonstrated an appreciation of the aesthetic and food value of Australian animals among Victorian acclimatizers. Recommending native animals for export, in conjunction with Mueller and Bennett's concerns about colonial environmental damage and waste, illustrates that Victorian acclimatizers were not inherently hostile to, or alienated from,

Colonial Creations 55

all native species. This position is in stark contrast to previous arguments that New Zealand acclimatizers positively disliked native animals or the linking of the entire movement to colonial alienation.[74] In fact, acclimatization was a wildlife management regime where utility, beauty, and local and imperial zoological knowledge governed which native animals to protect and what exotic creatures to import.

Importing exotic animals into Australia was shaped by local scientific priorities, particularly McCoy's acclimatization theory, but only when theory converged with increasing the colonial food supply, correcting, or compensating for environmental damage, and aiding in the colonization of new territories. Examining the animals recommended for importation within *Answers Furnished* was very revealing. It became apparent that acclimatization in Victoria was not about attempting to create a New Britain, proven by the fact that the majority of the organisms the ASV desired at this time were not native to Britain. Instead, the ASV desired to bring all (including appropriate British animals) climatically appropriate, useful "representative types" of animals from across the globe to Australia to improve the food supply and aid the processes of exploration and colonization. The detailed case studies investigated how these broader trends interacted with particular scientific traditions, the movement of ideas and people between colonies, and the practical realities of capturing and shipping viable populations of wild animals to Australia.

Thus far the ASV's attempt to and reasons for attempting to establish new fish species in Victoria has not been discussed. The next two chapters will address this omission and return the focus more closely on how the ASV understood and changed fish distribution in Victoria. Chapter 4 will look at the relationships between acclimatization, the ASV, commercial fishermen, and the regulation and exploitation of native fish. It looks at how the Victorian fisheries were understood in terms of taxonomy, environmental damage, and commercial potential. By doing this, it will reveal how the influence of European aquaculture, local observations, and legal precedents led the ASV into involvement with fisheries management.

CHAPTER FOUR

Regulating and Understanding Victorian Fisheries

> What a dreary land would this be to good old Isaac Walton! The race of anglers is in great danger of dying out in these colonies from sheer want of occupation. In the very best of times—the pastoral period of our history, before Mammon turned every river and creek into sludge channels—fish were anything but plentiful and the varieties were of a limited description.
>
> —*Yeoman and Australian Acclimatiser*, September 1863

The article above acknowledged the environmental damage caused by the gold rush, as well as the absence of familiar fish species in Victorian rivers. It proposed to improve and restore local rivers via aquaculture. Chapter 5 will explore colonial aquaculture and the ASV in considerable detail. First, however, I will investigate how the ASV and broader colonial society understood and valued local fisheries and fish species. These investigations will place the ASV's aquaculture program in context and demonstrate how Victorian acclimatization became as much about wildlife management as importing exotic species.

The ASV's fishery regulation policies were conditioned by particularities of local climate, taxonomic conventions, economic ambitions, environmental change, and emerging political traditions. These factors were in constant dialogue with imperial and transnational legal and scientific traditions in Europe, Britain, and North America. The ASV's fisheries management program was simultaneously an attempt to develop the commercial potential of Australian fisheries and an effort to protect and prevent damage to the fisheries developed in concert with the colonial government, commercial fishermen, and local scientists. To understand this process, it is necessary to look beyond the ASV's leadership and analyze the actions of local fishermen, the damage done to local fisheries during and after the gold rush, and legal precedents as well as to deepen our conception of the development of colonial science. Investigation will show how the ASV was reacting to and continuing actions originally initiated by commercial fishermen, who in turn built upon their experiences of declining fisheries and subsequent regulation in the United Kingdom.

Fisheries' management in early colonial Victoria must be seen in the light of the environmental, social, and economic effects of the Victorian gold rush. In Victoria, the gold rush, by causing unprecedented environmental damage, made some colonists aware of the dangers of overexploitation of forests, animals, and waterways.[1] Newspapers accused miners of being like locusts "who tear up and leave the earth desert in a few weeks" and making previously pure watercourses "thick and foul with gold washing."[2] Very few Victorian streams remained unaltered by the gold rush, and many suffered through flash flooding, erosion, and deforestation. These changes created an imperative to understand and protect Victorian fisheries that drew on European precedents, local and metropolitan taxonomies, and practical observations.

A keen observer of the Victorian commercial fisheries in the 1860s could have noticed commercial whaling on the southwest coast, small-scale seine net and line fishermen working Port Phillip Bay and Western Port to supply the domestic market, and Chinese fishermen fishing much the same areas but focused on salting fish for consumption on the goldfields and export to China.[3] As the railways pushed north, a small but rapidly increasing aquatic fishery supplemented marine harvesting. It targeted Murray cod and golden perch from the Murray Darling river system.[4] The fifth, and final, element was the dredging of oysters from Port Phillip Bay and Western Port for local consumption. There was also a small but avid group of amateur anglers fishing for river blackfish, grayling, mullet, Murray cod, and golden perch.

The Murray-Darling river system was exploited as a commercial fishery in a limited but significant way by the Murray River Fishing Company (MRFC). A failed American prospector, Joseph Rice, and six partners formed the company in 1855. It operated as a cooperative. With the help of Aboriginal fishermen, by 1869 they caught on average 170 ton of Murray cod and golden perch a year for the Bendigo and Melbourne markets.[5] Additionally, they harvested leeches, which were then exported overseas for medical use.[6] The ASV would both collaborate with and oppose the MRFC over the course of the 1860s.

The Yorta Yorta people laboring upon the Murray were not the only non-Europeans working in the Victorian fisheries. Chinese fishermen focused on catching fish for salting and drying. These Chinese merchants and fishermen were instrumental in developing the colonial fisheries.[7] In Victoria Chinese fishermen and fish curers operated at St Kilda, Phillip Island, and the Geelong lakes and around Wilson's Promontory. It has been estimated that starting in the late 1850s and continuing until the 1870s, Chinese fish curers accounted for over 76 percent of fish sales within the Australian colonies.[8] Sometimes as much as eleven tons of dried fish could be traded in a week between Victoria and New South Wales. Chinese fish sales did not go through the established colonial fish markets, which did not endear them to colonial authorities or other fishermen. Local fishermen argued that their Chinese competitors used nets with a very fine mesh, destroying immature fish that they thought unfit for market. This practice would bring Chinese fishermen into conflict with the ASV and the colonial government's fish protection policies as they developed in the mid-1860s.[9]

Early records indicate that at the time of first colonization (1835), the Yarra River and Port Phillip Bay held plentiful supplies of fish.[10] When Victoria's population exploded in the late 1850s, the local fisheries were exploited by local fishermen to increase the food supply. Before the gold rush only around twenty fishing boats operated in Port Phillip Bay and Western Port; by 1860 over 170 boats were operating in the same area.[11] Fortunately, it is possible to piece together the early development of the Victorian fisheries from newspaper accounts, government reports, and papers the ASV published.

One such paper was the "Fisheries of Victoria" (1864) by Samuel Lang, a former Scottish commercial fisherman who relocated to Victoria to pursue his trade. Lang attempted in 1841 to establish a deep-sea fishery company in Victoria, but local fishmongers' refusal to buy his fish and sabotage by independent fishermen drove the company out of business.[12] He still, however, maintained an ongoing interest in the fisheries and observed that by 1864 there were approximately 790 fishermen in Port Phillip Bay and Western Port operating 300 small boats.[13] These small-time fishermen primarily used fish traps and seine nets to ply their trade. Their target species included snapper, flathead, garfish, and rock cod. Approximately 250 imperial tons of snapper alone were being taken annually by the mid-1860s. Lang maintained that the coast and sea within easy reach of Melbourne contained a supply of fish that was "practically limitless."[14] This statement is eerily reminiscent of Thomas Huxley's concluding remarks at the British 1863 Fisheries Royal Commission that the act of fishing alone could not damage the supply of fish.[15]

Lang's optimism about the munificence of the Victorian fisheries, "most bounteously stocked by nature," was tempered by his feeling that the fisheries needed to be regulated, "as the wanton destruction now going on is most sinful."[16] Lang was not alone in calling for a regulated fishery. By 1862 the Victorian Angling Society (formed that year and collocated at the ASV's offices) was worried that colonists were in the process of destroying their rivers because "we sweep away our woods indiscriminately, and without a thought to to-morrow; we shoot down game in season and out of season, until many kinds threaten to become extinct; while we banish the fish which once enriched our coasts and rivers by wholesale processes of destruction. Some varieties, formerly abundant, are now scarce in the waters of Port Phillip."[17] It argued that the Victorian oyster fisheries, which "formerly supplied 12,000 to 14,000 oysters a year," were rapidly being destroyed.[18] Oysters were extensively dredged in Port Phillip Bay and Western Port in early colonial Victoria, and several companies existed for their commercial exploitation.[19] As early as 1856, calls were made to restrict the sale of oysters during certain months of the year, for sanitary purposes and to allow the oysters to spawn.[20] These calls built on a common belief that oysters were unsafe to eat when they spawned.[21] Just a year later, questions were asked in Parliament about the destruction of oysters within Western Port, and legislation was promised to regulate the industry.[22] In 1859 the Victorian Parliament passed the Act for the Protection of the Fisheries of Victoria, the Act for the Preservation of Fish in the Lakes and Rivers of the Colony of Victoria, and the Act for the Regulation of the Oyster Fisheries in Victoria. They were designed to control fishing techniques, suppress Chinese fishing, create a closed season for oysters, and license fishermen.[23]

These new regulations needed enforcement if they were to be effective. To that end James Putwain, a commercial oysterman from Sandridge, was appointed the first inspector of fisheries and oyster beds. During his tenure, he produced a report on the fisheries of Victoria that was reproduced in part in the colonial newspapers and in full in the *Public Lands Circular*, a monthly pamphlet distributed around country Victoria to explain the workings of the land acts.[24] Putwain's report reiterated the species targeted by the bay fishermen and the fishing methods they used. More importantly, he recorded and observed which species he believed spawned within Port Phillip Bay. In Putwain's estimation, bream (*Acanthopagrus butcheri*, or possibly *Acanthopagrus australis*) spawned in brackish water, silverfish (*Caranx georgianus*) spawned in shallow water over grassy beds, and the so-called Australian salmon (*Arripis trutta*) cast their spawn in shallow water.[25] McCoy provided an addendum to the report that listed Latin names of the fish species and promised to deliver a guide to Victorian fish shortly. Putwain and McCoy's observations represent the first attempt to catalog the breeding habits of fish in Port Phillip Bay and use this local knowledge to calibrate fisheries laws. Putwain further wrote: "I beg to state that there is a great diminution in several species of fish within the last six years, viz., whiting, mullet, guard fish and bream, which is attributable to the use of small nets. . . . I do most firmly believe, and am able to produce the opinions of some of the oldest fishermen in the colony, that if one-inch mesh be allowed to be used, it will be the means of entirely exterminating species of fish that have not as yet fallen victims to the small-meshed nets."[26] By the end of 1861, however, he was removed from office, officially to save money but possibly because of his opposition to the proposed relaxation of the net size regulations.

WHEN THE ASV FORMED IN 1861, it inherited growing local concerns about the state of the fisheries, increasing demands for fish, and emerging, but contested, local scientific conceptualizations of the morphology and habits of native fish. These factors in turn were conditioned by and affected growing international interest in declining fisheries and aquaculture. On a much broader level, the ASV's fisheries policies were conditioned by the three competing and interrelated concerns: developing the colonies' natural resources, repairing colonial environmental damage to maximize food production, and assessing and rectifying the perceived absences in the local economy of nature.

In April 1864, the ASV appointed a fisheries committee consisting of Edward Wilson, Professor McCoy, Lieutenant-Colonel Ross from the Angling Society, Samuel Bindon, and Herbert Watts. The committee was empowered to "collect and report upon all available information with reference to the varieties, the habits, the seasons and the qualities of our marketable fish with a view to their protection and increase and the consequent development of the fishery trade in this colony."[27] William Lockhart Morton, the inventor, pastoralist, and editor of the *Yeoman and Acclimatiser*, wrote that "of the really valuable fishes we know probably scarcely a quarter of those which exist in the Australian seas, and those which we do know and can get at we allow to be mishandled until they are threatened with utter extermination."[28] He went on to illustrate this process via the declining Western Port oyster fisheries, Murray cod, and Tasmanian seals. In the ASV's view, fish protection, classification, and the development of a commercially viable fishery were inextricably linked. Unfortunately, the fisheries committee never tabled its report on the native fisheries. It is possible to discern which species the ASV was interested in from which fish it thought would make valuable exports overseas, and what species it attempted to protect through legislation.

The ASV, and colonial Victorians more generally, valued specific native fish but viewed Australia as a piscatorial wasteland. In 1859, while drumming up support for acclimatization, Edward Wilson wrote a series of articles in the *Times* titled "The Distribution of Animals." In one of these articles, Wilson wrote that after observing anglers in the Yarra, "I always felt grieved to think that they had not the slightest chance of catching anything of above a pound weight; the only fish in the river, except eels, being a little 'blackfish,' rarely reaching that weight, and a fish we call 'herring,' which does not attain anything like the size of its saltwater namesake."[29] In the 1864 *Answers Furnished* questionnaire, McCoy and Mueller wrote that except Murray cod and golden and silver perch, Australian fish "are not of great size, and perhaps not equal to many European kinds in flavour."[30] In a later paper, McCoy attributed local enthusiasm for fish acclimatization to the fact that "the species of fish good for the table are very much fewer in Victoria than in Europe; and great interest attaches, therefore, with many of the general public, to the endeavours of the Acclimatization [sic] Society of Victoria to introduce the salmon and other good British fish into the waters of the colony, independently of the scientific interest of the experiment."[31]

The Victorian colonists were not entirely wrong. Australia is relatively poor in freshwater fish species, and those that do exist are significantly different from European fish. Even today there are only 256 native fish species recognized in Australia, as opposed to 360 in Europe and 5,000 in South and Central America.[32] Australian freshwater fish either descend from early Gondwanan species or have evolved recently from marine fish.[33]

Strange did not necessarily mean unloved. Murray cod are Australia's largest purely aquatic fish; the biggest recorded specimen weighed in at 114 kilograms.[34] Cod are long lived, predatory, and nonmigratory. They occur naturally in many rivers and lakes along the southeast coast of Australia. Before white invasion, Murray cod was a major food source for Aboriginal peoples living on the Murray-Darling river system. In some Aboriginal mythologies, a giant Murray cod is responsible for creating the Murray-Darling river system.[35] The ASV could not stop praising the Murray cod, considering it a "first-class fish of the table," and extolled both its flavor and its virtues as an angling fish.[36] The ASV's enthusiasm for Murray cod led it to attempt to introduce cod to lakes and rivers in Victoria where they did not naturally occur. Unsuccessful attempts were also made to establish Murray cod in England, New Zealand, and France.

Regulating the Murray cod fisheries posed a problem for the ASV and the colonial governments because the cod fishery was based on the Murray River, which was deemed to be part of New South Wales. The fishermen were based in Victoria and supplied the Bendigo and Melbourne markets. Over three successive years (1863–65), the ASV was informed that in the Murray a "great destruction was going on amongst the Murray cod."[37] The ASV's unease with the overexploitation of the Murray cod fishery was part of a broader undercurrent of disquiet with the MRFC's practices. In September 1864 a letter was written to the Echuca newspaper the *Riverine Herald* accusing the MRFC of harvesting cod, during the breeding season, that were full of roe.[38] A letter to the *Pastoral Times* succinctly argued, "The fishing company do not consider that taking fish when they are spawning is slaying the goose that lays the golden eggs."[39] Concern for declining fisheries was also framed in terms of sympathy for the local Yorta Yorta people; critics accused the MRFC's fishing practices of "affecting our sable friends, the aborigines who complain that they cannot get fish in season."[40] In fact, some Yorta Yorta people made complaints about the destruction of their fisheries to the Central Board Appointed to Watch over the Interests of the Aborigines.[41]

In 1863 the ASV offered to cooperate with the MRFC to put a stop to the wanton destruction of Murray cod.[42] The MRFC replied that only "a very small destruction of the Murray Cod took place."[43] In the local press, the MRFC argued "they desire protection as much as anyone can do; that it is against their interests to destroy fish in the breeding season, and they do not believe they are getting scarce."[44] It shifted blame for the destruction of cod to the Yorta Yorta people, arguing, "The blacks, it seems, are very destructive of the fish. It is the peculiar privilege of their women to catch the young fry, which they do in nets, securing thousands in a day. It is no uncommon thing in the season to see a lubra [Aboriginal woman] trudging home of an evening with a load of these young fish, under which she can scarcely move. In former years, when the blacks were much more numerous, this work of destruction was carried on to a far greater extent that it now is."[45] Dissatisfied with this response, and informed that wholesale destruction was continuing in the Murray, the ASV wrote to the NSW government and the ASNSW about the continuing destruction of cod.[46] The ASNSW replied promising action.[47] New South Wales did not initiate effective inland fisheries regulations until after the 1880 Royal Commission Enquiring into the Fisheries of New South Wales, and even then enforcement was sporadic.[48] The ASV did not break ties with the MRFC despite its role in the destruction of cod. In 1866 the ASV ordered 150 young cod from the MRFC for export to England and France.[49] The native fish that the ASV exported and the fish species it valued were affected by taxonomic conventions and the development of colonial ichthyology.

Eric Rolls once wrote that Australia was "more like a new planet than a new continent."[50] He was attempting to describe how early colonists perceived Australian flora and fauna, particularly marsupials, as strange and alien. Australian rivers and lakes, and the fish within them, evoked different feelings. Colonial fishermen and scientists alike were struck by how much some native fish resembled familiar European and North American fish in taste, smell, and morphology; simultaneously they were perplexed by the absence of familiar fish species and indeed the absence of entire fish families.[51] Scientists and fishermen looked to taxonomic ichthyology and their own practical experience to explain Australian fish distributions.

The taxonomy of fish was still in its infancy in the mid-nineteenth century. It was fragmentary, and taxonomists were unable to agree on classifications or species names. This slowly improved as large-scale taxonomic works were published. Works and authors of particular importance to Australian ichthyology include Georges Cuvier's and Achille Valenciennes's multivolume *Histoire Naturelle des Poissons* and Albert Günther's ever-expanding *Catalogue of the Fishes in the British Museum*.[52] Australian scientists, including McCoy, George Bennett, and Gerard Krefft, contributed to these more substantial works by supplying specimens and publishing scholarly articles that shaped how Australian fish were perceived.[53]

Controversies over taxonomic relationships were possible because of disputed definitions of the salmonid family and a tendency to record minor regional variations as separate species.[54] In the words of John Gray, keeper of the zoological section of the British Museum, salmonids "offer so many and so great difficulties to the Ichthyologist, that as much patience and time are required for the investigation of a single species as in other fishes for that of a whole family. . . . Sometimes forms are met with so peculiar and so constantly characterized, that no ichthyologist who has seen them will deny them specific rank; but in numerous other cases one is much tempted to ask whether we have not to deal with a family which, being one of the most recent creation, is composed of forms not yet specifically differentiated."[55] Taken as a whole, taxonomic ichthyology slowly bound Australian fish into a tangled web of kinship that could be interpreted in multiple ways by scientists and citizens.

In colonial Victoria, the local galaxia fish were known colloquially as "river, or Yarra trout," and *Prototroctes maraena* were commonly known as the Yarra grayling or Yarra herring. Both colloquial and scientific taxonomies asserted kinship between these fish and salmonids.[56] When writing about the possibility of acclimatizing salmon in the Yarra, fisherman and ASV Council member H. E. Watts argued that the "Australian grayling is almost identical with the English fish in character, in habits and in the manner of capture"; he further argued that the wide distribution of grayling in Victorian rivers "is interesting to our acclimatizers, as indicating the fitness of our rivers to receive other members of the salmon family. Where the grayling lives it is tolerably certain that the trout will live too."[57] The hunter and amateur naturalist William Wheelwright contested Watts's assertion, arguing that Australian grayling were not true grayling and that the Yarra was an unsuitable place to acclimatize salmon and trout.[58]

Cuvier established the galaxia genus and asserted its kinship to Salmonidae. Cuvier's student Valenciennes scientifically described the first Australian species of galaxia, the *Galaxias truttaceus*, or spotted mountain trout, in 1846 in volume eighteen of their coauthored *Histoire Naturelle des Poissons*.[59] Valenciennes followed Cuvier and classified *Galaxias truttaceus* as part of the salmonid family.[60] Krefft accepted Cuvier and Valenciennes's judgment and asserted that galaxias are "a tribe of fresh-water fishes representing the trouts in the southern hemisphere."[61]

Albert Günther, the curator of the British Museum's piscine collections and eminent British ichthyologist, argued that Australia's freshwater fish were split between two biogeographical regions, the equatorial and the southern regions.[62] He believed both areas were deficient in fish species due, in the first place, to the arid climate and deficiency of water in the Australian continent.[63] Additionally, Günther believed that there were two particular families of fish in the southern biogeographical region, each of which "is analogous to a northern type, viz the *Haplochitonidae*, which represent the *Salmonidae* (Haplochiton being the analogue of Salmo, and *Prototroctes* that of *Coregonus*), and the *Galaxiidae*, which are the pikes of the southern hemisphere."[64]

McCoy accepted Günther's taxonomy and used his authority to reinforce Watts's colloquial assertion of affinity between Australian and European grayling. He rejected the idea that galaxias were close taxonomic kin with trout. He wrote to Watts stating that the anatomy of the Australian and European grayling was very similar.[65] Furthermore, McCoy pointed out that Günther had recently described the Australian grayling as "having the general habit of Coregonous" and creating a new family for it that included the trout-like Haplochiton of South America.[66]

The disputed relationships between native fish and salmonids affected the ASV's approach to fisheries protection (and helped determine which species the ASV acclimatized). Fishermen and the ASV saw Australian rivers refracted through a vision of European salmon streams, projecting onto Australian fish, not always accurately, ideas of annual migrations and set breeding seasons. The strangeness of Australian fish did not alienate Australian colonists; instead, they were tantalized by the similarities between European and native fish and worried that Australian fisheries might crash like their European counterparts.

The ASV's legislative preferences must be seen in the context of fisheries legislation in Britain, and, to a certain extent, France. In the early 1860s, Britain began a long and arduous process of standardizing riverine and estuarine fishery restrictions. The 1861 Salmon Act was the first attempt at comprehensively regulating the declining salmon fisheries in England. The Salmon Act grew out of the 1861 Royal Commission Appointed to Inquire into the Salmon Fisheries (England and Ireland), whose report recommended the establishment of enforced closed season, the prohibition of set nets, banning the establishment of new weirs, and the appointment of professional fisheries inspectors.[67] The 1861 Salmon Act prohibited set nets and the establishment of new weirs without adequate fish ladders; it did not, however, provide adequate enforcement measures for the new restrictions. Further legislation was called for, and the 1865 Salmon Act passed. This act required licenses to be issued for fishing nets, creating local conservation boards, and appointing water bailiffs to enforce the laws.[68]

Putwain and the ASV sought to imitate the British salmon acts and prevent local fisheries from declining. The ASV sought to defend Putwain when he was dismissed. He was invited to attend one of the ASV's fortnightly meetings where he read his 1861 report on the state of the fisheries and emphasized the need to maintain one-and-a-half-inch net meshes to protect spawning fish.[69] At this same meeting, the ASV resolved to write to the newspapers about fisheries protection and to approach the commissioner of customs about Putwain's dismissal.[70] George Sprigg, the ASV's secretary, wrote to the *Argus*:

> The deterioration and apparently early destruction of many fishing-grounds in the United Kingdom and in France having excited a good deal of attention, the council of the Acclimatisation Society wish me to write a few lines to you, informing you of the danger which threatens our fisheries, and the measures which, with the co-operation of the Legislature, they hope to be able to adopt to effect their preservation. In so doing, I shall avail myself largely of information derived from a very valuable report recently laid before the society [sic], from the late inspector of fisheries.[71]

The ASV's reaction to Putwain's report and his dismissal is significant for several reasons. The ASV situated its concerns about declining fish stocks in Victoria among debates concerning declining fish stocks in Europe, showing the ASV's determination that metropolitan fishery declines not be repeated in Victoria. Small declines in Victorian fish catches triggered fears of more substantial future declines, based on English and French precedents. In listening to Putwain's report, the ASV demonstrated the importance they placed upon observations of local conditions, fish habits, and fishing methods. On a much larger scale, the ASV came to believe that to develop commercial fisheries in Victoria, it was necessary to regulate net sizes, types, and positioning.

In pursuing a regulated fishing industry, the ASV had the support of some, but not all, commercial fishermen who operated in Port Phillip Bay and Western Port. The ASV sent a delegation, including Thomas Black and McCoy, to the commissioner of customs to petition for Putwain's reinstatement, the prohibition of set nets, and the preservation of the one-inch net mesh.[72] The commissioner stated that Putwain was dismissed to save money, and promised to consult the ASV before introducing a new fisheries bill.

The 1862 Act to Amend and Consolidate the Laws for the Protection of the Fisheries of Victoria incorporated most of the provisions of the earlier acts and implemented further restrictions recommended by the ASV. In common with the earlier acts, the 1862 Protection of the Fisheries Act protected fish introduced into Victorian rivers and lakes for two years, prohibited fishing using poison, required the registration of fishing boats, and made it illegal to sell fish caught illegally.[73] Significantly, the Protection of the Fisheries Act preserved seasonally differentiated net mesh restrictions (one and a half inches in winter and one and three quarters inches in summer), as advocated by the ASV and opposed by some commercial fishermen.[74] Seasonal net mesh restrictions echo provisions within the 1861 British Salmon Act and accord with Putwain's observations of seasonal migration patterns among some Victorian estuarine fish species.[75] This demonstrates the central importance of both metropolitan precedent and local observations to the ASV and to Victorian fisheries legislation. The 1862 Protection of the Fisheries Act did contain some new provisions that were not included in the earlier acts. Fixed nets were prohibited within one mile of shore and were banned outright at the mouth of any river.[76]

Neither the ASV nor those fishermen opposed to regulation saw the 1862 Protection of the Fisheries Act as an end to their struggles. On 14 August 1863, Charles Nicholson tabled a petition from "certain fishermen" in Parliament asking that the governor in council be given the power to allow fishing with nets in "that part of the River Yarra Yarra extending from its junction with the Saltwater River [Maribyrnong] to its junction with the waters of Hobson's Bay."[77] James Meek, who was both an indifferently successful fisherman and an artist and cartographer of some local renown, appealed directly to the ASV, asking that they allow the winter net width restrictions to be reduced to one inch in order allow the successful harvesting of guard-fish in Port Phillip Bay.[78] While the ASV could not directly change fisheries regulation, Meek's appeal to them indicates that the ASV was perceived as a critical participant in colonial fisheries' regulation. Dr. Black referred Meek's letter to Lieutenant-Colonel Ross and the Angling Society. Unfortunately, a complete transcript of Meek's letter has not survived. Ross's reply extensively quoted it and stated, "The Angling Society, having considered the tenor of Mr Meek's letter, is of the opinion that if the proposal therein contained were granted (which he must know no society has the power to grant) there would not be many guard-fish left in the bay of Port Phillip in the course of a very few years."[79] Ross furthermore argued that decreasing the net width would indiscriminately destroy young and old fish, effectively reintroducing what he called the "Chinese system" of fishing, which previous legislation banned and "from the effects of which the fisheries of Port Philip Bay have not yet recovered."[80] It was, in Ross's opinion, foolish and self-defeating for fishermen to advocate for indiscriminate harvesting practices. He thought that the "total destruction of the oyster-beds at Western Port might be a very useful lesson upon this head."[81] His reply to Meeks indicated an awareness of earlier Victorian destructive fishing practices, a continued animus toward Chinese fishermen, and an increasing but contested knowledge of the habits of local fish species.

The ASV appointed Ross to lead a delegation to the commissioner of customs tasked with discussing problems with the 1862 Protection of the Fisheries Act. It took the position "that the fisheries would never be well protected until the Government appointed a properly qualified inspector."[82] Additionally, it argued that the 1862 Protection of the Fisheries Act poorly defined the prohibited fishing zones near estuaries, making it hard to prosecute offenders. It also wanted the protections over introduced fish to be extended and increased. In September Ross met with the commissioner of customs to discuss problems with the Protection of the Fisheries Act. The commissioner asked Ross to compile a list of problems with the act and send it to him, which Ross did.[83] In April 1864 the Act to Consolidate the Laws for the Protection of Fisheries and Game passed, combining the fisheries and game laws. The

1864 consolidation act allowed the governor in council to extend the protected period for translocated fish and to extend protection to other rivers; these were both regulations suggested by the ASV.[84]

The ASV continued to debate proper regulation of the fisheries with local commercial fishermen. McCoy and experienced local fisherman William Webber were involved in a dispute concerning the best way to conserve fish in Corio Bay, near Geelong. Webber wrote to McCoy and the ASV that the firing of artillery near Corio Bay had damaged the local fishery.[85] McCoy rejected this assertion out of hand, instead attributing the decline to Chinese fishermen destroying young fry in the bay. Webber and the other Corio fishermen disagreed. Drawing on their experience in Corio Bay and managing commercial fisheries in England, they maintained that the destruction of young fish would have no effect on adult populations, "as the porpoises and other large fish eaters would destroy the young fish if the Chinese fishermen did not destroy them."[86] Webber, furthermore, vehemently maintained that while he was a fisherman in Cornwall, his "own personal knowledge" led him to observe how the Falmouth Castle battery could disrupt shoaling fish.[87] McCoy also drew on British precedents, arguing that laws restricting net sizes "have not only been made in Britain, on the recommendation of fishery commissioners having ample evidence to guide them, but, further, that there has been time to recognise a revival of the old abundance of fish in many localities since the enactment of laws for the protection of the young fish."[88] McCoy and the ASV prevailed, and Victorian legislation was not loosened.

Some Victorian fishermen and legislators continued to see regulation as an improper restriction on free trade and their livelihood. One fisherman, calling himself "an advocate for small pilchards if I cannot get large ones," wrote a frustrated letter to the *Argus*, arguing:

> The Deep Sea Fishing Company are promising great things for the poor of Victoria, by supplying trawl-fish at 1½d. per pound. I am sure they need it. They cannot afford to pay 7d. and 8d. per pound for beef and mutton—the price we have to pay at Williamstown. I cannot agree with the gentleman from St. Mawes [Webber], who stated the other day that it was a sin to catch so small a fish. I say a greater sin would be to let them go, while beef and mutton are so high, and poverty and starvation staring poor people in the face.[89]

The ASV continued to advocate against the "sin of catching small fish" and for the long-term sustainability of the fishing industry. In 1867 Thomas Black delivered a talk on British fisheries legislation to the ASV's monthly meeting, admiring the appointment of professional inspectors and attempts to prevent the destruction of immature fish.[90] The ASV frequently reported violators of the fisheries act to the police and offered small rewards for successful prosecutions under the fisheries act.[91] In 1869 it sought an amendment to the fisheries act allowing the destruction of birds that preyed on young fish.[92] The organization also unsuccessfully lobbied for restrictions on the size of fish that could be caught.

After the ASV lobbied for further restrictions, and a failed fisheries bill, the Victorian Parliament commissioned a board of inquiry to prepare an amended fisheries bill. The inquiry solicited testimony from professional fishermen, fish salesmen, and other interested parties. During the inquiry, seine and set net fishermen argued about which form of fishing was more destructive as well as the perennial topic of net-mesh sizes. As a way around this problem, it was suggested, with the support of the ASV, that net mesh restrictions be replaced with prescribed legal weights for each fish species. Also, the board recommended the appointment of professional inspectors of the fisheries, another cause dear to the ASV. Both professional inspectors and prescribed legal weights were included in the 1873 Act to Amend the Law Relating to Fisheries. It was a bittersweet victory for the ASV, because after the passing of the 1873 fisheries act, the ASV's role in fisheries policy passed to professionals.

FOR MANY REASONS, the mid-1870s proved to be a transitional point for the ASV and the Victorian fisheries. As mentioned, professional fishery inspectors were created and slowly appointed as part of the 1873 fisheries act, decreasing the ASV's central role in the formation of fisheries policy. The ASV itself was, by the 1870s, in the process of transforming into a zoological organization with declining interest in acclimatization. New regional fish acclimatization societies, such as the Geelong and Western District Fish Acclimatising Society and the Ballarat Fish Acclimatisation Society, were founded and acted as independent bodies not subject to the ASV's central authority. Private individuals, such as Samuel Wilson, also began extensive fish acclimatization programs independent of the ASV.

Fisheries regulations in Victoria were shaped by the adaptation of European regulatory practices and ichthyology conventions to Australian circumstances by commercial fishermen, the ASV, and colonial scientists. These groups aimed to understand local fisheries to conserve commercially valuable native species and prevent the repeat of European-style overexploitation and fishery collapse. Victorian fishermen bequeathed to the ASV an awareness of the vulnerability of and damage done to the local fisheries. The ASV responded to this knowledge by studying the distribution and morphology of local fish and the potential of local rivers to harbor new fish species. The ASV concluded that the Victorian fisheries were commercially valuable though damaged and that there was a taxonomic and functional affinity between local fish and salmonids. Building on these assumptions, the ASV attempted to regulate local river fisheries as if they were British salmon streams, with minimal success.

The significance of this analysis is both specific and wide-reaching. On the smaller scale, it has been demonstrated how the ASV valued native fish, were aware of environmental damage to the Victorian fisheries, and made critical contributions to multiple fisheries acts. These actions show a more complete vision of acclimatization in Victoria, one that does not allow the ASV to be seen only as environmental vandals, dedicated solely to introducing exotic species. On a broader scale, the analysis provided further evidence to support recent contentions that colonial Victorians were aware of and concerned about the environmental consequences of the gold rush.[93] It has also further illustrated the networked and contingent nature of colonial science. Looking at fisheries management, however, reveals only part of how the ASV and colonial Victorians understood environmental change and biogeographical distribution. To complete the picture and illustrate how the ASV attempted to restore and improve Victorian fisheries, chapter 5 will explore colonial aquaculture.

CHAPTER FIVE

Aquaculture

In the 1880s a series of books were published that detailed the introduction of salmonids into Victoria and Tasmania. These portrayed the process as a deliberate and planned attempt to re-create British angling in the Australian colonies. These early accounts of salmonid introduction have colored scholarly understanding of acclimatization, colonial aquaculture, and colonial environmental history more broadly.[1] Recent studies of aquaculture and fisheries management in North America and Europe suggest that aquaculture has a more complicated history than simply re-creating and extending recreational angling. Furthermore, the 1880s accounts of Australian aquaculture and the scholarship that has used them are incompatible with the arguments developed in the previous four chapters. While there is no consensus regarding the development of salmonid aquaculture in North America, several scholars have pointed out some factors that are salient when studying acclimatization in Victoria. First, there is an ongoing tension between recreational and commercial fishing. Second, the scholarship points out a reoccurring argument that rivers supporting relatively few species should be stocked with species from elsewhere. The third and final recurring factor was the idea that damaged fisheries can be restored not necessarily with the original species but with similar but tougher species from elsewhere that are more capable of surviving in degraded environments.[2] The ASV's aquaculture program demonstrates all these elements.

The program became entwined in a complex and reciprocal exchange with the ideas and practices of the American environmentalist George Perkins Marsh. He was interested in aquaculture and fish acclimatization as a way of restoring damaged fisheries. Furthermore, the Victorian salmon acclimatization program reinforced Marsh's interest in aquaculture. Marsh, in turn, inspired American fisheries scientists like Livingstone Stone.[3] There are definite parallels between fish importations conducted by Livingstone Stone in California in the 1870s and the ASV's fish acclimatization program in the 1860s, including being motivated by a dissatisfaction with the distribution and qualities of native fish, as well the degradation of the fisheries.[4]

Victorian acclimatizers theorized about environmental change and helped create acclimatization in the British Empire, welding it to local priorities and problems. Similarly, British aquaculture was a coproduction of Britain and its colonies. Chapter 2 demonstrated that acclimatization science in Victoria was theoretically distinct and focused on introducing organisms deemed to be taxonomically kin to Australian animals, both to improve economic production and to restore colonial environmental damage. Chapter 4 brought to light perceived taxonomic kinship between Australian native fish and salmonids, and local fisheries under stress from overfishing. This entire book demonstrates continuing tension between acclimatization as ecological imperialism and acclimatization as neo-ecological imperialism.

The ASV's aquaculture program did not result from the actions of a group of conservative men alienated from and unable to comprehend Australian nature. Instead, the program emerged from a combination of factors: the ASV's observations of the degradation of local fisheries, engagement with and reworking of the latest European and British ichthyology and aquaculture research, and the desire to create and expand commercial fisheries. This process can be observed within the multiple failed attempts to introduce Atlantic salmon (*Salmo salar*) to Victoria and Tasmania and the resulting transition to trout acclimatization.[5] These experiments were dependent on and contributed to the emerging science of aquaculture.

MODERN AQUACULTURE WAS BORN in the late eighteenth century. From the very start, it was concerned with restoring damaged fisheries, while simultaneously developing their commercial potential. The precise origin of artificial fish breeding is perhaps best credited to a series of late-eighteenth-century naturalists and landowners. In 1763 Ludwig Jacobi, a German army officer, published an article in the *Hanover Magazine* that detailed a procedure for artificially incubating trout eggs, with specific reference to incubation times and the effect of temperature on their development.[6] In the late 1840s three different Frenchmen, including two French farmers, Anton Géhin and Joseph Remy, and a prominent savant and Société Zoologique d'Acclimatation member Armand de Quatrefages, published papers on the artificial raising of fish. Quatrefages claimed that fish culture was capable of compensating for all the current causes of fish destruction in France.[7] Meanwhile, in Britain, Gottlieb Boccius emphasized the uses of aquaculture in *A Treatise on the Production and Management of Fish in Freshwater by Artificial Breeding and Rearing*.[8]

Nevertheless, France commissioned the first state-sponsored scientific investigation into the state of the fisheries and the plausibility of using aquaculture to restore damaged environments. These studies were conducted by Professor Victor Coste and led to French imperial sponsorship for the creation of artificial oyster beds and the establishment of an extensive fish breeding and experimentation station at Huningue in Alsace-Lorraine.[9] The Huningue "piscifactoire" (literally, fish factory) developed techniques for raising many species of freshwater fish from eggs (including trout, salmon, and perch) and successfully transporting them to rivers. James Youl would later visit Coste's Huningue facility when attempting to develop techniques for transporting trout and salmon ova to Tasmania.[10] European imitators of Huningue abounded all over Europe.

The Ashworth brothers and their industrious employee Robert Ramsbottom were crucial figures in the development of British aquaculture. Indeed, they attempted to replicate the success of Huningue. Edmund and Thomas Ashworth were Quaker merchants from Manchester who leased salmon rivers in Ireland. Robert Ramsbottom's background is mostly unknown, other than that he was a professional fishing tackle manufacturer and an acquaintance and later employee of the Ashworths. He did all the practical work involved in artificially breeding salmon for the brothers. In 1866 Ramsbottom published a booklet titled *The Salmon and Its Artificial Propagation*. The brothers themselves published several books on the subject, including *A Treatise on the Propagation of Salmon and Other Fish* (mainly a translation of Coste's *Instructions Pratiques sur la Pisciculture*), and an *Essay on the Practical Cultivation of a Salmon Fishery* (1866). In these works, the brothers detailed how they had leased the Galway fisheries, and through a process of artificial breeding, the removal of set nets in rivers, and the appointment of water bailiffs, had increased the yield of salmon harvested at the river mouth.[11] Like Coste in France, they saw rivers as "water-farms," and proposed mechanisms for transporting salmon ova across vast distances. Importantly for salmon acclimatization in Australia, the Ashworth brothers were interested in the commercial potential of salmon fishing, not angling.

Ramsbottom and the Ashworths also helped establish the Stormontfield salmon breeding establishment on the Tay River in Scotland. In 1853, after being informed about the Ashworths' experiments in Galway, the proprietors of the Tay River salmon fishery resolved to begin trials in the artificial breeding of salmon.[12] The Stormontfield site was selected, and Ramsbottom initiated several decades of artificial stocking and salmon-breeding experiments. These experiments produced a wealth of knowledge about salmon breeding and inspired a generation of eager aquaculturalists. Experts agreed, as discussed in chapter 4, that by the mid-nineteenth century Atlantic salmon stocks were in decline in Europe. Furthermore, and essential to the acclimatization of salmon in Victoria, French and British ichthyologists figured out the development time of salmon ova and how to slow down development by lowering the temperature, thus holding out the tantalizing prospect of establishing commercially viable Atlantic salmon fisheries in the Australian colonies.

A NEXUS OF BRITISH and colonial individuals, dedicated to exporting commercial salmon aquaculture, formed in Britain and the Australian colonies. In Tasmania and Victoria, the advocates for salmon acclimatization were a rough coalition of acclimatizers, agricultural reformers, and pastoralists aided by British experts who provided personnel, expertise, fish, and support. This alliance, formed in 1859 with a shipment of ova on the *Sarah Curling*, was solidified during and after the *Beautiful Star* shipment and broke apart as agendas diverged in the aftermath of the *Norfolk* shipment. However, before we explore this, much can be learned by an exploration of the *Columbus* shipment.

The very first attempt to introduce salmon to Australia occurred before the formation of this nexus. In 1849–50, at the initiative of James Ludovick Burnett, an official in the Tasmanian Survey Department, 50,000 salmon ova were placed aboard the *Columbus*. Andrew Young, the aquaculturalist, author, and manager of the Duke of Sutherland's salmon fishery, offered advice on how to proceed. Nevertheless, the attempt failed when the fertilized ova died en route.[13] The Tasmanian government then sought the help of Gottlieb Boccius, an established British aquaculturalist. Boccius advocated extensive aquaculture of salmon, trout, and carp; strictly controlled fishing seasons; the removal of weirs from rivers; and closing salmon streams to angling. He, like Coste and Marsh, believed that fisheries were declining and that aquaculture would lead to "a greatly-augmented increase of employment, a new source of revenue, and a large supply of wholesome food, at a price so reasonable as to bring it within the means of the poorest classes."[14] This sentiment informed the subsequent attempts to acclimatize Atlantic salmon in Tasmania and Victoria.

In 1859, thirty thousand salmon ova, collected by Robert Ramsbottom, were placed upon the clipper *Sarah Curling* and shipped to Tasmania. The shipment was funded by the Tasmanian government and was organized principally by Edward Wilson and James Youl. The British aquaculturalist Alexander Black accompanied the ova. Despite the fact that the ova were destined for Tasmania, it is relevant to acclimatization of salmon in Victoria because it involved prominent Victorians and illustrated the broader ethos of the early salmon experiments in the Australian colonies.

JAMES YOUL WAS A PROMINENT and successful Tasmanian pastoralist, who left Tasmania in 1854 but retained an active interest in colonial matters throughout his life.[15] He has long been acknowledged as a central figure in acclimatizing salmon in Tasmania, but very little research has been done on his motivations for introducing salmon. In 1860, while the *Sarah Curling* ova were still in transit, Youl wrote a letter from London to the editor of the *Launceston Examiner* urging colonial governments to commit funds to salmon acclimatization. He considered salmon fisheries to be more profitable in the long term than gold mining. He wrote: "Just consider what exertion would be used to discover a gold-field that would yield annually as much as the Irish Salmon Fisheries, £350,000; and there is this very great advantage in the one over the other—that the goldfield becomes exhausted of its riches, whilst with care your salmon fisheries, instead of becoming exhausted, will increase year by year in value."[16] Youl saw the establishment of salmon fisheries as a commercial proposition, not as a source of recreation. He drew the £350,000 per annum figure mentioned from information supplied to him by Thomas Ashworth.[17]

Edward Wilson supported the *Sarah Curling* shipment to demonstrate the plausibility of the acclimatization of European fish in Australia and to popularize it in both Britain and the colonies. To understand what species to ship to Australia and how to do so, Wilson corresponded with the Ashworth brothers. Thomas Ashworth referred him to *A Treatise on the Propagation of Salmon and Other Fish* and recommended not acclimatizing salmon but instead concentrating on trout, carp, and tench, because these were hardier species. Despite this advice, Wilson drew his own inference from the Ashworth brothers' books that salmon might potentially be acclimatized in Victoria and Tasmania.

Frank Buckland explicitly drew on Wilson and the Ashworth brothers to justify fish acclimatization. He referred to a (no longer extant) letter he received from Wilson about "utilising the waters" in which Wilson argued that "fish hatching will pay," and referred to the profits of the Scottish and Irish salmon fisheries. Also, Buckland quoted Thomas Ashworth's report on the Huningue fish-breeding establishment, stating what fish it bred and how it would serve the French national interest.[18] Buckland's speech helps illustrate how salmonid aquaculture and acclimatization were a coproduction of Britain and the Australasian colonies.

This coproduction included an extensive report Black wrote to the Tasmanian Parliament on the potential benefits of salmon aquaculture after the *Sarah Curling* ova died. He detailed the expected profit margin and growth rate of salmon in Tasmanian waters.[19] He expected that given the introduction of 300,000 salmon fry per annum into Tasmanian rivers for six years that the rivers may "be considered adequately stocked, and capable of furnishing an inexhaustible supply of fish."[20] Black anticipated that the salmon would be commercially harvested within three years of their introduction to Tasmania and would net those responsible a tidy profit. He furthermore advocated the restriction of recreational salmon fishing in Tasmania until the species established itself, combined with the assisted migration of experienced commercial salmon fishermen to Tasmania to harvest the fish.

It was not mere optimism that led Black to believe that salmon would thrive in Tasmanian waters. He conducted an extensive survey of the temperatures of Tasmanian rivers and bays, comparing them with the temperatures in established salmon fisheries in Scotland and Ireland. Drawing on previous research, Black maintained that salmon will not spawn in water more than fifty-five degrees Fahrenheit and that the temperature of the principal salmon rivers of Scotland and Ireland varied between thirty-eight and forty-six degrees Fahrenheit during the year.[21] By comparison, he noted that the Derwent River's average temperature was forty-four degrees Fahrenheit and that the seas around Tasmania were of similar temperature to the English Channel. These similarities, along with a lack of predatory fish in Tasmanian rivers, led Black to conclude that "the rivers of the colony are suitable for a nursery and habitation for Salmon, the adjacent marine feeding grounds will afford abundance of food, and the temperature of the water prove congenial to the fish."[22] Black's report indicated that salmon could survive in Tasmania, that they were potentially profitable, and that it was worth making another attempt to acclimatize Atlantic salmon.

Based on Black's assessment, and on advice supplied by Thomas Brady, the secretary of the Irish Fisheries Board, the *Beautiful Star* ova shipment proceeded on 4 March 1862. Unlike earlier voyages, this attempt at salmon acclimatization was supervised by the newly appointed Tasmanian salmon commissioners and was jointly funded by the ASV.[23] The ASV financed the project in spite of the fact that establishing salmon in Tasmania (and not

Victoria) was the project's aim. The ova were, this time, accompanied by William Ramsbottom, the son of Robert Ramsbottom. An elaborate mechanism consisting of glass rods, multiple layers of boxes, and many water pumps was used to transport the ova, but to no avail. Most of the ova died in transport, except for 300 ova that were packed in peat moss and sealed for the entirety of the journey.

The continual failure of Atlantic salmon ova to survive the trip to Australia and mounting skepticism of their ability to thrive in Victorian rivers forced the ASV to reassess and systematize its aquaculture program. Simultaneously, as argued in chapter 4, the ASV was struggling to come to grips with overexploited native fish populations and wondering whether small-scale local declines prefigured European-style fisheries collapse. In 1864 the ASV created a subcommittee consisting of prominent local scientists and fishermen to organize their fish acclimatization program. The committee concluded "the salmon is the most desirable of the fishes" but also that the ASV should "procure the ova of the following fishes, trout, salmon trout, especially also the grayling, perch, and char."[24] All of these fish, bar the English perch, are members of the salmonid family and were held to be climatically and anatomically suitable for acclimatization in Victoria. The ASV; its successors, the regional fish acclimatization societies; and wealthy private individuals spent decades attempting to acclimatize all of these species with varying levels of success.[25] The ASV persisted in trying to acclimatize salmonids despite early failures. There were a number of reasons for this. First, as discussed in chapter 4, there were the perceived taxonomic similarities between native fish and salmonids. Second, there was practical knowledge about how to breed salmonids. Finally, it was recognized that salmonids possessed commercial potential and had the potential to supplement local native fish supplies, which were under threat from commercial fishing and environmental degradation.

The *Yeoman and Australian Acclimatiser* continued to advocate importing Atlantic salmon in spite of the failure of the *Beautiful Star* shipment. It drew on both Buckland and the Ashworth brothers to demonstrate the commercial potential of Atlantic salmon fisheries. It quoted Buckland as saying, "Gold nuggets (in the shape of valuable fish and the possibility of rearing them) have long been under our noses in the water, and we have not stooped to pick them up."[26] The paper reinforced this message by pointing out that the Ashworth brothers believed "it is cheaper and easier to breed salmon than lambs," and that salmon aquaculture could improve the Australian population's commerce and diet.

Acclimatisation Society Council member and *Argus* co-owner Lauchlan Mackinnon argued that "the Salmonidae have a tolerably wide diffusion over all the waters of the temperate zone, and are common to all the countries of Northern Europe, America and Asia. Some of the genera have been found in tropical waters; and at least two entirely distinct and new ones have very recently been shown to inhabit our Australian waters, one of them being the so called fresh-water herring of the Yarra, and the other a sea fish sometimes found in Hobson's Bay."[27] The sea fish Mackinnon was referring to was most likely the "sea trout" (*Arripis trutta*), a species harvested by colonial fishermen in Victoria. The fishermen held them to be a form of salmon because of their pink flesh and superficial morphological resemblance to Atlantic salmon. The other fish was the Australian grayling (*Prototroctes maraena*), which had a purported kinship with salmonids.

Mackinnon's article justified salmonid acclimatization using McCoy's acclimatization theory.[28] He and other colonists maintained that there was a "gap in the economy of nature" with regard to fish in Victoria, fulfilling the first criterion of McCoy's theory. McCoy's opinion that Australian grayling were representative types related to the Salmonidae fulfilled the second criterion, the presence of representative types. The third criterion, climatic suitability, was fulfilled when the salmon commissioners in Tasmania and the ASV itself conducted studies showing the suitability of salmonids for Victorian and Tasmanian waters.[29] Given that the salmonids fitted McCoy's criteria so well, it is unsurprising that the committee reported to the ASV's council that "the salmon is the most desirable of the fishes."[30] In 1867, McCoy himself wrote that the Australian grayling's "close resemblance in food and habits to the true Salmonidae helped the Acclimatization [sic] Society to argue that certain of our rivers would serve for the experiment of acclimatizing [sic] the European salmon and trout."[31]

Building on this enthusiasm and knowledge, on 15 April 1864 approximately 100,000 salmon ova and 3,000 trout ova arrived in Melbourne aboard the *Norfolk*. The vast majority of the ova were immediately transferred to Tasmania, and only a small proportion were retained in Victoria for small-scale experiments. Soon after the arrival of the *Norfolk*, the relationship between the ASV and the Tasmanian salmon commissioners soured. In May 1864 the salmon commissioner Morton Allport read a paper before the Royal Society of Tasmania detailing how the *Norfolk* shipment of salmon and trout ova was transported to Tasmania and Victoria but also complaining that all the ova should have been transferred to Tasmania because none of the Victorian salmon had any chance of survival. The ASV took umbrage at this assertion, arguing that it had a prearranged agreement with the Tasmanian salmon commissioners to share the ova, calling Allport "ungracious and churlish."[32] Furthermore, the ASV maintained that it was its duty to attempt the acclimatization of salmon in Victoria.[33]

When the salmon ova arrived in Melbourne, the ASV established a salmon subcommittee consisting of Mueller, McCoy, Madden, and Black.[34] It was responsible for the care of the ova and young fry and selecting a site for releasing the salmon. Several sites were considered and rejected, including the Snowy River and several rivers in Gippsland.[35] It was eventually decided to raise the young salmon fry in a suspended tank in Badger Creek, a tributary of the Yarra River. According to the ASV's correspondence, the salmon thrived in Badger Creek from 1864 to 1865. In May 1865 the ASV released thirty, seven-inch salmon, from Badger Creek, into the Yarra River. None of these fish were verifiably seen again.[36]

Meanwhile, the Tasmanian proportion of the *Norfolk*'s salmon and trout ova was hatched and placed in specially built lakes to encourage growth before eventual release. The salmon suffered considerable mortality. The trout, on the other hand, thrived in the ponds and soon began to spawn.[37] A representative of the ASV visited the Tasmanian salmon ponds and was not impressed. He reported that the salmon ponds were partially stagnant and unsuitable for fish accustomed to fast-moving clear streams, recommending that either the ASV redesign the ponds or the salmon immediately be released into local rivers. The salmon commissioners released the Tasmanian salmon into the Derwent River in 1865, and despite numerous sightings, not a single fish taken from the Derwent River was incontrovertibly identified as an Atlantic salmon.[38]

IN SPITE OF THE UNDERWHELMING success of the *Norfolk* salmon, the ASV and the salmon commissioners together arranged another shipment of ova. This time the load was placed aboard the *Lincolnshire* and consisted of 102,500 salmon ova and 15,500 trout ova. In spite of jointly funding the operation, the ASV decided to retain only a small proportion of the trout ova and to forward all of the salmon ova to Tasmania.[39] This represented a definitive shift in policy. Unfortunately, the ASV's papers make no explicit reference as to its decision to cease Atlantic salmon acclimatization. However, it is plausible that the failure of the *Norfolk* salmon to become established played a significant part in its decision. By 1868 the ASV had decided to focus on nonmigratory salmonids, specifically brown trout, to maximize the chance of success and minimize losses to poaching and migration.[40]

A. C. Cooke, *Salmon Tanks in the Badger Creek, Upper Yarra*. *Australian News for Home Readers*, 1865. Courtesy of the State Library Victoria.

The shift away from salmon aquaculture and toward trout acclimatization was partially unintentional—trout bred well in captivity; salmon did not. Within two years of their introduction to Tasmania, brown trout were spawning both within the salmon commissioners' ponds and in local rivers. In fact, by 1864 brown trout were regarded as "established in our rivers beyond all risk of failure."[41] Rapidly establishing breeding populations of trout in Tasmania was important to Victoria because it meant the ASV could quickly and cheaply acquire trout ova for acclimatization experiments, unlike salmon ova, which had to be transported laboriously from Europe. Easy access to ova allowed multiple acclimatization experiments to be conducted in Victoria, lessening the cost of failure and leading to eventual success. After several failures (including the *Lincolnshire* trout ova), trout were hatched and liberated in Riddells Creek outside of Melbourne in 1868.[42]

Success with trout acclimatization did not represent an ideological shift within the ASV or among the salmon commissioners. Both groups remained committed to aquaculture as a means of feeding the population and seeding commercial fisheries. The salmon commissioners continued to argue that "when the almost incalculable value of the Salmon and Trout—as articles of human food, as a means of extending our commerce, increasing our population, and affording employment to our labouring classes—are considered, besides the direct pecuniary returns to the Treasury of the Colony, the expense incurred in their establishment in our rivers sinks into significance."[43] These arguments were a continuation of the economic justifications for salmonid importation used since the beginning of the Tasmanian project. The ASV also saw trout acclimatization as part of its broader plan of improvement and restoration. George Sprigg, the ASV's secretary, heartily endorsed a proposal to stock newly established reservoirs with trout:

> The art of pisciculture, although practiced centuries ago, has only recently been revived, and already it may be said to have worked miracles in some of the exhausted fisheries of Europe.
>
> How important it is for us, in a climate like that of Victoria, and with comparatively but a limited supply of fish, to endeavour to utilise to such a purpose the magnificent series of reservoirs which will shortly be in the possession of this colony. By inaugurating a complete system of fish-breeding a most important addition may be made to the food of the people whilst at the same time the public revenue will benefit largely by the result.[44]

Sprigg's endorsement demonstrates clear continuities within the ASV and the salmon commissioners' ambitions for Atlantic salmon in Australia and a broader imprint of European aquaculture's commercial and restorative aspirations. The *Argus* went further when arguing for a comprehensive aquaculture program in Victoria. It pointed out that "Marsh, in his *Man and Nature*, believes that in every populous country, all barren and worn-out soils, wherever the situation admits, will yet be converted into lakes for the propagation of fish, and that the quantity of food thus obtained will diminish the constant necessity for enlarging the area of cultivated land."[45] To the *Argus*, this argument meant that Victoria would be wise to invest in an extensive aquaculture: "Our lakes, fresh and salt, will then be of more value acre by acre than our more solid public domains," supplying food for rich and poor alike.[46]

Ideology and fecund fish from Tasmania were not sufficient to establish brown trout in Victoria and feed the population. The ASV's Royal Park fish-breeding facility was inadequate. To counter this problem, in 1874, the ASV was given a special supplementary grant of £300 specifically to establish a trout-breeding facility. Curzon Allport (the brother of Morton Allport) supervised its construction. He was previously involved in the successful breeding ponds in Tasmania, where the design of the Stormontfield fish-breeding facilities in Britain informed his practices.[47] The Wooling site was designed to create an independent breeding population of trout in Victoria, separate from the Salmon Ponds in Tasmania. In spite of raising and distributing some trout, the Wooling facility never really achieved its aim due to uncertain water supply, poor maintenance, and the depredations of cormorants and poachers.[48] By 1880 the Zoological and Acclimatisation Society of Victoria (ZASV; the ASV's name after 1973) had abandoned attempting to breed trout at Wooling and switched its efforts to encouraging private fish-breeding facilities. The ASV's attempts at trout breeding were not responsible for establishing these fish across the state. Its program was an unintentional side effect of its attempt to develop a commercial salmon industry and increase the local food supply.

THIS CHAPTER, IN CONCERT WITH CHAPTER 4, has revealed a more nuanced view of fisheries regulation and aquaculture in 1860s Victoria that is at odds with older, totalizing conceptualizations of Australian environmental history. Chapter 4 established that local scientists and fishermen saw parallels between the anatomy of native fish and salmonids. They also saw similarities between the overexploitation of native fish in Australia and salmon in Europe, leading them to pass fisheries legislation similar to the British Salmon Acts and to perceive Australian waterways as akin to British salmon rivers. Given this analysis and recent aquaculture scholarship, this chapter reexamined aquaculture in 1860s Victoria, focusing on Atlantic salmon and brown trout. European aquaculture provided the ASV and the Tasmanian salmon commissioners with personnel, knowledge, and most importantly, precedents for using aquaculture to establish and restore commercial fisheries. Local expertise and taxonomic conventions that saw a kinship between native fish and salmonids, combined with European technology that enabled the stripping and transfer of salmonids, led the ASV to focus on acclimatizing them in Victoria. The first importations of brown trout were merely by-products of attempts to establish a commercial salmon industry; as will be discussed in chapter 8, recreational angling became significant only in retrospect. This reexamination of aquaculture in Victoria reinforces the overarching argument that acclimatization in Victoria rested on complex and astute observations of the local environment, some elements of which acclimatizers valued highly, combined with transnational scientific knowledge and commercial imperatives.

The next chapter moves on from fish and fisheries toward analyzing how acclimatization in Victoria encouraged, understood, and regulated hunting. It will look at successive Victorian game acts and their enforcement. While the officers of the ASV played a vital role in these processes and will be subject to an in-depth analysis, to truly understand acclimatization in Victoria it is necessary to move beyond the ASV and look at secondary actors. In the case of hunting, the critical secondary actors are commercial and recreational hunters, scientists, the state, and farmers. The interactions between these groups and the organisms they, or the ASV, imported or valued were refracted through reactions for and against British hunting traditions, declining native wildlife, and scientific understandings of acclimatization and biogeography.

CHAPTER SIX

Hunting Victoria

> No man, here at any rate, need be so hopelessly poor or so socially low as to feel himself excluded from the fullest participation in the successes which may attend the efforts of our acclimatisation societies.
>
> . . . for the sooner suitable animals can be introduced in large numbers the sooner will his table become the recipient of venison, shot, it may be by his own hand in the wild wood, to roam in which he will have an undoubted right. To the advantage of varying his diet he may add the pleasures of the chase, a fair glimmering expectation of which has already made its appearance in the successful liberation, in various parts of the colony, of Ceylon elk and axis deer.
>
> —*Ballarat Star*, November 1863

This extract was part of an article that argued for further acclimatization and the diversification of the Australian diet. It contains many of the critical elements that were emergent in colonial Australian hunting practices, including deeply ambivalent relationships with British hunting traditions; the blurring of the line that demarcated subsistence hunting from hunting for sport; and the involvement of the acclimatization societies. The formation of Australian colonial hunting traditions and the ASV's complicated relationship with agricultural development, environmental degradation, and imperial recreational activities helps explain this quote and further reveal the complexities of acclimatization practices in Victoria. The evolution of the ASV's relationship with hunting practices and environmental attitudes can be drawn out by an investigation of the creation, development, and enforcement of the Victorian colonial game acts in the 1860s and early 1870s.

Scholars have paid only minimal attention to the history of game hunting in Victoria, and the game laws themselves have been misunderstood. Some have argued that the statutes reflected a continuation of abhorrent English traditions and that they ignored the protection of native animals.[1] It would be surprising, given the analysis of other acclimatization practices, if hunting fitted into this older model, not least because it does not reflect the broader scholarship that has developed in the last three decades about colonial empires of hunting.

John M. MacKenzie helped establish colonial hunting as a sphere of scholarly endeavor. He argued that ritualized British imperial hunting practices slowly supplanted native hunting regimes in India and Africa, serving to reinforce and display British imperial power.[2] Several studies have built upon MacKenzie's work, and the idea of native subsistence hunting being slowly replaced by sports hunting as the empire expanded has become something of a historical orthodoxy.[3] Recent studies, however, have shown that in Canada, Tasmania, and parts of India, subsistence and commercial hunting existed side by side and that subsistence hunting was never entirely subsumed by sports hunting.[4] Sports hunting had regional variations and developed during the nineteenth century. In New Zealand, deer hunting evolved to focus on deer stalking, as opposed to organized battues or mounted hunting parties. While deer hunting required a game license, these licenses were relatively cheap and allowed many middle-class families to supplement their diet with venison.[5] Instead of one monolithic imperial game narrative, hunting evolved a multiplicity of meanings and practices, dependent on geography, faunal distributions, and local traditions.

Chapters 4 and 5 argued that the ASV played a crucial role in advocating for the protection of Victorian fisheries and attempted to introduce fish species to expand the commercial fisheries and increase the food supply. The ASV's relationship with farmers and fishermen grew more complex and antagonistic as alpacas and salmon failed, and sparrows became agricultural pests. It is inadequate to see acclimatization as resulting purely from the ASV's initiative. Secondary actors were vital to establishing, legitimizing, and contesting acclimatization in Victoria. Furthermore, acclimatization cannot be seen only as based solely on the importation of exotic species. Indeed, it encompassed the regulation of native and introduced animals in Victoria. Neither were Victorian acclimatizers deferent to or seeking to reproduce Britain. Nor is it helpful to treat Britain as a stable point of comparison in contrast to the dynamic colonies; both engaged in ever-renegotiating webs of meaning and action.

Four different and interrelated groups negotiated the establishment, enforcement, and evolution of the Victorian game acts: the ASV, the organized colonial hunt clubs, commercial game hunters, and farmers. These groups argued about environmental damage and the correct use of native and introduced animals. They sometimes collaborated but were often in conflict because of differing interpretations of the value of British hunting traditions and game laws. Out of this contested space emerged Victorian game laws and hunting traditions that were distinct from but shaped by British laws.

Australian attitudes toward game and hunting were conditioned by game laws and hunting practices in the United Kingdom. The game laws in Britain evolved piecemeal the thousand years after the Norman Conquest but became codified in the late seventeenth century.[6] It is, however, mainly the eighteenth- and nineteenth-century forms of and attitudes toward hunting that affected debate in Victoria. The nineteenth century saw the escalation and industrialization of British hunting practices. The development of breech-loading shotguns enabled hares and game birds to be shot more efficiently, which led to increased competition over bag size, and the development of battue hunting, which involved large-scale hunts where birds were driven toward stationary hunters. To produce the amount of game required for hunts of this scale, it was necessary to employ an army of gamekeepers, to lock up vast swathes of land as game reserves, and to breed game artificially. All this served to heighten hostility between hunters and tenant farmers. There has been a long historical debate concerning poaching in Britain during this period and whether it should be best considered a form of rural protest, class war, or organized crime.[7] Game preservers became more organized, using an army of gamekeepers and deploying mantraps and spring-loaded fixed shotguns to deter poachers. Carrying the equipment necessary for poaching (such as nets, guns, and hunting dogs) became punishable by law. Punishment for poaching became ever more severe (codified in the 1828 Night Poaching Act) and included transportation to the Australian colonies.[8] Poachers, in turn, became more organized, militant, and violent.[9]

By the 1830s it became apparent to some British legislators that the current game laws were obtuse, contradictory, and ineffective. In 1831 a new game act was passed that legalized the sale of game, and replaced property qualifications with an expensive game license (a license cost £3 13s 6d). The legal market in game was supposed to undermine the market for poached game, but in reality, decreased prices led to a higher volume of poached game that could be sold legally with less risk to the poacher.[10] The 1831 game act was amended in the House of Lords, reintroducing transportation as a punishment for poaching and making game unambiguously the property of the landowner, rather than the occupier of the land.[11] This act had several pernicious effects for tenant farmers: they could not destroy pest animals that harmed their crops (rabbits, hares, pheasants), they were legally held accountable for any poaching on their land, and they could be compelled to create conditions favorable for game on their leased properties.

Hunting Victoria

In 1845 John Bright MP, one of the leaders of the Anti-Corn Law League (who campaigned against laws that restricted the importation of grain), convened a parliamentary inquiry into the game laws that addressed farmers' concerns. He took evidence from gamekeepers, poachers, and tenant farmers. Poachers argued in their testimony that it was the threat of starvation or the workhouse, coupled with the belief that wild animals were a common resource, that drove them to poach.[12] During the course of his inquiries, Bright deduced that game animals destroyed crops, further inflating grain prices.[13] He found that tenant farmers saw this destruction as a violation of their property rights. Conversely, great lords and property owners considered defending the game laws as integral to defending their class and customary status. Bright's inquiry had very few practical outcomes, but it began a discussion on the effect of game on agriculture, and the property rights of tenant farmers.

These issues reemerged in Britain in the late 1850s and early 1860s, the same period when the need for game laws was being discussed in Victoria and the other British colonies. By 1859 wheat prices were declining and a rabbit plague had broken out over parts of England.[14] Throughout the 1860s, farmers began to agitate for the right to destroy rabbits, for protection of their property rights, and for revision of the game laws. The British game laws bequeathed to the Australian colonies both a model for game preservation and active resistance to game laws as infringements on the property rights of farmers, as well as representing the excessive regulation of resources that should be freely available to all.

In Victoria, British game and hunting practices and legislative traditions were contested and transformed by several interrelated groups practicing different forms of hunting. Interested parties included commercial wildfowl and kangaroo hunters, the ASV, organized hunt clubs, squatters, urban radicals, and a newly emerging class of small farmers. Colonial Victorians practiced four different kinds of hunting: commercial hunting for the urban markets, subsistence hunting, vermin elimination (kangaroos, dingoes, and emus), and ritualized sports hunting. The line between these different forms of hunting was often blurred: sportsmen turned commercial hunters, small farmers sometimes participated in organized hunts, and hunt clubs targeted kangaroos and dingoes before the establishment of deer and foxes in the colony.

Hunting culture in Victoria cannot be separated from attempts to establish and expand agriculture within the colony. Native animals, such as emus, kangaroos, or dingoes that competed for food with or preyed on agricultural animals were declared vermin and hunted. They were pursued on horseback with the aid of kangaroo-dogs, which were greyhounds cross-bred with bloodhounds that resembled the traditional poachers' Lurcher, both for fun and to protect crops and livestock. The French traveler and property speculator Hubert De Castella described a typical hunt in *Les Squatters Australiens*. He wrote of kangaroos as "far too common" and complained that he had approximately 1,000 of them on his land and that they consumed as much fodder as 150 sheep.[15] In 1854 he and four friends went kangaroo hunting along the Yarra River with the aid of some "big greyhounds." He praised the speed of the kangaroos, the determination of the hounds, and the sheer joy to be had from pursuing kangaroos at a gallop.[16] Author, politician, and land reformer William Westgarth also sang the praises of kangaroo hunting, in 1848: "The spirit and eagerness of the dogs, and the loud 'whoop' of the huntsman, as he discerns his prey coursing with rapid and fantastic gait through the open forest, excite incredible ardour and enthusiasm. Nothing can be more inspiring in this description of pastime, so long as the bounding tenant of the woods maintains himself ahead of his pursuers. But the successes of the sport are attended with a touching spectacle, in the innocent and piteous expression of the suffering victim."[17] These authors, by praising the hunters and the virtues of kangaroos, partially transformed kangaroos from vermin to highly desired and admired game animals, akin to deer or foxes. The transformation was only ever partial, and by 1864 Westgarth was noting with pleasure efforts to exterminate kangaroos and lamenting that, due to declining dingo and Aboriginal populations, in some areas kangaroo numbers had risen.[18] He was also aware (and disapproved) of a general decline in native animals around Melbourne. The rough hunting practices described by De Castella and Westgarth venerated the victim of the hunt similar to stag and foxhunting in Britain. Unlike in Britain, hunting in Victoria was unregulated; anyone who could afford to maintain a horse and a dog could participate. Without further ritualization, hunting could not be used to distinguish master from man, a large squatter from a small farmer, and a farmer from the digger.

Recent migrants from England went straight to the goldfields, where they learned to treat the local environment as both a resource to be exploited and common property to be shared.[19] Hunting was a crucial part of this process as miners hunted both for pleasure and for food. Unrestricted access to hunting was a symbolic break with British tradition and a demonstration of the liberty afforded to miners.[20] Some noted with regret the massive erosion, declining wildlife numbers, and deforestation caused by mining.[21] The gold rush thus bequeathed Victoria with the dual legacy of open communal access to resources and the consequences of unbridled exploitation.

Many miners later became small farmers. Throughout the 1850s and 1860s there were extensive campaigns to unlock the lands, break up the squatters' leaseholds, and establish yeoman agriculture. Most of the ASV's leadership was in favor of this process. Throughout the 1860s, successive land acts were passed aiming to establish small-scale farming in the colony. Squatters exploited loopholes in these acts to secure the land they already leased.[22] Selectors (as the farmers who purchased freehold blocks from the government were known) aimed to assert their property rights and respectability. Many selectors struggled to make their farms viable.[23] Selectors were simultaneously vulnerable to native and introduced animals damaging their crops, as well as needing to supplement their diet through hunting.[24]

Large squatters (land holders who leased vast pastoral properties from the government), on the other hand, aimed to distinguish themselves socially and economically from selectors and miners by establishing grand houses and affecting the mannerisms and habits of British gentry and nobility. One of the ways they did this was to establish formal hunt clubs and transform the rough hunting of the early colonial period into an ersatz copy of British fox hunting, minus the foxes. These clubs were elite organizations that allowed members to socialize, network, and through knowledge of hunting etiquette, demonstrate gentlemanly conduct. The Melbourne Hunt Club traces its origin to 1853, when George Watson imported some hunting hounds from his father's pack in Ireland.[25] His father was the master of hounds in County Carlow. Several other packs were established in Corio and Werribee but were subsumed into the Melbourne Hunt. Watson established himself as a land and stock agent and was soon styling himself "Master of the Melbourne Hunt."[26] Like their more informal counterparts, the hunt clubs targeted dingoes, emus, and kangaroos. Later they imported fallow deer and foxes. Unlike the informal hunts, the hunt clubs maintained membership lists, used pedigree fox hounds, and had pretensions to exclusivity. Police Magistrate Foster Fyans's description of the Colac squatters encapsulated the hunt clubs' aspirations. He observed the aftermath of an organized dingo hunt, "among this class of gentlemen squatters in no instance did any man exceed, or forget that he was a gentleman."[27] The line, however, between gentlemen and working men, between hunting for pleasure and hunting for food, was fine in early Victoria. Squatters destroyed vermin, farmers followed the hounds, and down-on-their-luck British sportsmen turned to commercial hunting to make ends meet.

One such British sportsman-turned-commercial-hunter, William Wheelwright, recorded and published his activities as *Natural History Sketches — By the Old Bushman*. His account is the most complete source of information concerning commercial hunting in Victoria. He spent two years as a kangaroo hunter and several more observing commercial duck hunting in the colony. When kangaroo hunting, Wheelwright shot and snared over 2,000 kangaroos in the Western Port District for their hides and meat.[28] He observed that kangaroos were becoming rare in the settled areas, and thought that the colonists' practice of killing kangaroos on sight and leaving the carcass to rot "does, indeed, seem a shameful waste of one of the bounties of nature."[29] Coursing for kangaroos with dogs was, according to Wheelwright, both inefficient and unsporting. After doe kangaroos were run down with dogs, their joeys were torn from their pouches and dashed against a tree, a practice that Wheelwright thought "with the exception of clubbing seals this certainly did appear to be about the most barbarous work I ever joined in."[30]

Wheelwright was even more skeptical of the practices of commercial duck hunters operating in the colony. "When I first came into this country," he wrote, "the palmy days of the duck-shooter were in their zenith; the fowls and buyers plentiful, the shooters scarce."[31] Within a year commercial hunters were exploiting wildfowl surrounding Melbourne. These hunters were catering to a demand for fresh meat fueled by the population boom caused by the gold rush and a scarcity of food caused by farmers abandoning their properties.[32] Increased demand had dire consequences for local wildlife and hunters because "as the birds became scarcer, the shooters increased, and prices fell, till at the present day duck-shooting is not worth following within fifteen miles of Melbourne."[33] Contemplating declining wildlife populations, Wheelwright argued for a local game act, which might create a closed season for game but did not license hunters or restrict who could hunt.[34] He claimed that when "the game of any country becomes a marketable article, and of sufficient value to induce men to devote their whole time to its pursuit as a means of gaining a livelihood, it should in some measure be protected, *especially as it is not private property* [author's italics]."[35] Wheelwright's views revealed an emerging and distinct Victorian approach to game management, built around public ownership of game and seasonally restricted access. Its foundations lay in the different hunting traditions discussed but also created conflict between different types of hunters and even with farmers.

ALL FORMS OF HUNTING PLACED PRESSURE on local game populations, leading to calls for the regulation of hunting and the importation of game animals from overseas. To address these concerns, local Victorian parliamentarians, with the support of the ASV, sought to regulate and codify local

Hunting Victoria 91

hunting practices, protect native game during the perceived breeding period, and protect newly imported animals to allow them to become established. The first Victorian game regulations were initiated in 1858 but failed to pass both houses of Parliament.[36] Four years later, the 1862 Act to Provide for the Preservation of Imported Game and during the Breeding Season of Native Game passed both houses and became law.[37]

Between the failed 1858 game bill and the successful 1862 game act, the politics of and reasons for game preservation were debated within the colonial newspapers. In 1860 the liberal Ballarat newspaper the *Star* argued that "a very large and unprofitable destruction takes place of what may be termed our Australian game" and that the perpetual destruction that goes on was both "a violation of all sportsmanlike rules and feelings, and a useless waste of the gifts of nature."[38] The *Star* argued that it was possible to regulate the shooting of game without replicating the British game laws, "which we most heartily denounce as iniquitous and largely productive of crime," so long as game remained public property.[39] As an example of this approach, the *Star* quoted a Californian game law that restricted shooting game during a closed season but did not make game private property. In Tasmania, newspapers treated a proposal to protect black swans with considerable skepticism. The *Launceston Examiner* questioned, "Of what special use are black swans to the colonists of Tasmania that the legislature should be asked to throw over them the shield of its protection? Neither the sportsmen nor the epicurean would give a pin for a gross of them."[40] The *Examiner* further argued that the primitive animals of Tasmania would go the way of its indigenous people and claimed that Tasmania will still exist, and perchance prosper too, "when the black swan, like the Dodo, will live only in the history of the past."[41] Contrastingly the *Cornwall Chronicle* lamented the extinction of native animals and thought that regulation was necessary.[42] However, like the *Star* in Victoria, the *Cornwall Chronicle* argued that the population would not tolerate "the passing of any acts similar to the British Game Laws" because of lingering and justified resentment of these laws in the colony.[43] In spite of some colonists' reservations, the Tasmanian Parliament easily passed the Black Swan Protection Act and Native Game Act in 1860.[44]

To fully understand the 1862 game act, and how it expressed such contradictory attitudes toward hunting, the political composition of the Victorian Parliament and the colony more generally need to be explained. The Legislative Assembly consisted of a range of groups who attempted to form a government through a variety of successive coalitions. These included outright conservatives, land reformers and their radical sympathizers, and moderates.[45] Land reform was a vital issue of the day. In 1860, a mutilated land act was passed that pleased no one and was widely viewed to have failed. Class antagonisms ran high. Charles Don, the radical member for Collingwood, threatened "to whip the squatters across the Murray."[46] Peter Snodgrass, the conservative member for Dalhousie, through the Victoria Association helped collect £30,000 to protect squatters and "support the rights of property."[47] These men had fundamentally divergent views on the proper use of property and natural resources.

The Victorian 1862 game act passed quickly but not without opposition. Snodgrass introduced the 1862 game act into the Legislative Assembly.[48] The 1862 game act is best understood as a marriage of convenience between sports hunters, individuals concerned with regulating commercial hunting, and the ASV. It is similar to the alliance made between commercial and recreational fishermen to control the local fisheries. The act intended to regulate when game could be shot, while still preserving game as public property open to all. By asserting closed seasons, the 1862 game act rejected the gold-digger tradition of completely open access to game.

The 1862 game act was cosponsored by William Jones, who explicitly stated that he did not want to introduce English-style game laws in Victoria, but "simply to protect birds brought into the colony by the Acclimatisation Society, which birds would be beneficial to the colony" and to protect useful native game birds.[49] The radical Member of Parliament Charles Don opposed the 1862 game act on several grounds.[50] He wished to preserve open access to game for the poor, expressed continuing class-based resentment of the British game laws, and saw the act as a further ploy by squatters to secure their leaseholds.[51] He feared that he was witnessing the rebirth of the hated English game laws, and mobilized traditional rhetoric of common ownership to resist them.

Don was only partially correct in equating the British and Victorian game laws. Examining the 1862 game act clause by clause demonstrates how it diverged from English legislation. The first clause defined "game" as a select group of introduced species and defined "native game" as a subgroup of valued native species. The second clause stated that game animals were to be protected for three years from 1861 to 1864, and native game protected during what was deemed to be the breeding season—1 August until 30 November.[52] These clauses did not create private property in game or restrict access to hunting to a particular class. Like the British game acts, the 1862 game act took a punitive approach to violators of the law and levied heavy fines for destroying game and native game out of season.[53] Clause six made it illegal to "buy sell or knowingly have in his possession" restricted game and native game. This clause was very similar to a provision that John Bright and the 1845 British Select Committee on the Game Laws unsuccessfully attempted to have inserted into the British game acts, indicating some level of continuity between the mid-nineteenth-century radical approach to game and the 1862 game act. Clause eight was problematic for the ASV because it exempted any game or native game in "confinement or in a domesticated state" from protection under the act. Thus, it effectively created private property in some forms of game.[54]

The ASV was hostile to the idea of game as private property. Its interest in game preservation can be traced to 1861, when it received multiple reports that wildfowl and quail were wantonly destroyed throughout the colony, thus alerting them to the possibility of declining bird numbers.[55] Consequently, the organization involved itself in the formulation of the 1862 game act.[56] These consultations established a pattern that would last for decades of the government using the ASV as a board of experts when it came to zoological matters. The government was still consulting the ASV on animal protection at the end of the nineteenth century.[57]

The minutes of the ASV's meetings indicate that it formed a subcommittee to consider the act and made recommendations to the commissioner of customs regarding the composition of the legislation.[58] The details of the ASV's proposals have not survived. However, the organizing principles of its interest in the game acts can be inferred by reference to an editorial written in the *Argus* by ASV Council member Lachlan Mackinnon with regard to the subsequent 1864 game legislation. Mackinnon wrote that the Victorian game laws were not designed to create private property in game, reserved for one class, but rather:

> Our object being simply to protect *game* and not the *proprietors of game*, we have nothing to do with any questions affecting property or class. We have to preserve the animals themselves, for the benefit of all — to guard them in the seasons when they require guarding, to take care that they are not wantonly destroyed, and to secure them every chance of reproduction.
>
> . . . The end proposed is to make game plentiful and cheap, and the pursuit of it open to all; and this cannot be done without some restrictions on individual greed, or individual thoughtlessness.[59]

Such views suggest an interest in "game animals" that goes beyond the sports hunting to concerns about the food supply and the recreation of both the rich and the poor.

The most problematic and intriguing elements of the 1862 game act, and the ones that most demonstrate the ASV's intervention, are schedules one and two (reproduced below). The first schedule lists game species in need of protection that the ASV had imported or intended to import. Some of these were also introduced independently by sportsmen, for example, hares and deer. Additionally, hares, deer, and pheasants began to be seen as pests by British tenant farmers over the nineteenth century.[60] These conflicting claims to ownership and control led to conflict among the ASV, farmers, and sportsmen with regard to imported game animals. The second schedule provides a list of the protected native game including species of interest to the ASV, commercial hunters, and sportsmen.

THE FIRST SCHEDULE

Pheasants £5
Partridges £5
Grouse £5
Hares £5
Deer £15
Swans £5
Antelope £15
All birds not indigenous to
    Australia and their produce £1

THE SECOND SCHEDULE

Wild Ducks of any species
Native Companions
Teal
Wild Turkey or Bustard
Bitten
Black Swans
Emu
Wild Geese
Bronzed-winged and other
Wild Pigeons
Mallee Hens[61]

Perhaps one of the most noticeable aspects of these schedules is the native and introduced species that do not appear in them. Kangaroos and dingoes were both targets of ritualized sports hunting and opportunistic culling by squatters and farmers.[62] Like foxes in the United Kingdom, these species were simultaneously "game animals" and, in the minds of farmers, vermin that killed lambs and competed for forage with sheep. Other species that were more unambiguously "game" were also excluded from the schedules. Native quail, snipe, and, most surprisingly, rabbits failed to make the lists. By 1862, Thomas Austin and others had successfully established rabbits on the Geelong coast and were hosting large-scale rabbit hunts.[63] Rabbits were not protected by the act for several reasons. First, it is possible that because rabbits were, at this early stage, largely confined to private property, they were considered to be "under domestication" and thus adequately protected as the property of the leaseholder. Second, it is possible that the Victorian legislation was following the precedent set by the United Kingdom, which asserted that while rabbits were the target of game hunting, they had no specific protected legal status.[64] Third, it is possible that by 1862 rabbits were already considered established enough not to be in need of legal protection. Certainly by the late 1860s, rabbits, like dingoes and kangaroos, were bitterly resented by some farming communities.

The native animals protected in schedule two and the imported species protected in schedule one includes species of long-term interest both to the ASV and to sportsmen, who together with farmers defined how to value species in 1860s Victoria. Up to now, the historiographical consensus has indicated that the ASV took little interest in native animals, or appreciated them only for their use as a means of exchange for desired exotic species. The ASV, and especially George Bennett, wanted to protect some native species and develop their agricultural and industrial potential. The ASV was interested in improving and safeguarding local fisheries. The ASV's interest in "game" animals must be seen in a similar light.

The ASV's interest in game laws was initially sparked by concerns about the wanton destruction of wildfowl on the North Melbourne and Albert Park swamps. Wheelwright's *Natural History Sketches* contains a detailed account of his commercial wildfowl hunting operations, Melbourne's gastronomic demand for wildfowl, and his concerns that wildfowl were in decline. Specific species mentioned by Wheelwright appear in the 1862 schedule two as protected native game. Among the wildfowl, teal and magpie geese were singled out for hunting because of their flavor and the consequent high price they fetched in the Melbourne market. The bronze-winged pigeon was another valuable commercial bird that Wheelwright thought was declining. All in all, he concluded, "it cannot be denied that the game is rapidly disappearing in all the settled districts, especially near town, and if steps are not speedily taken to prevent the wholesale destruction of the birds in the breeding season which is now carried on, in a few years the shooter's occupation in Victoria will be gone."[65] His complete "game list of Victoria," animals he thought required protection, included brush turkeys, native ducks, pigeons, quail, snipe, and rail. Like the initiators of the 1862 game act, and the ASV itself, Wheelwright did not want to replicate the British game laws or to create private property in wild animals. He wrote: "Let us have no licence. Let a man still be free to wander where he will on land that is not purchased but let us have proper seasons fixed for killing the game."[66]

Having explored two out of the three elements of the alliance of convenience that led to the passing of the game laws—sportsmen and commercial hunters—it is now necessary to turn our attention to the specific contribution made by the ASV to the two schedules and their enforcement. The ASV was consulted regarding the content of the act and the animals that should be protected. The precise nature of the advice it gave cannot be directly identified because the content of these consultations has not survived. The ASV's continuing interest in the native animals listed for protection within schedule two can, however, be extrapolated from its published annual reports, speeches, and other publications, as well as its attempts to see schedule two enforced once it was promulgated.

Looking back at chapter 3, it can be seen that the ASV recommended every native species protected by the 1862 game act for export overseas, and did indeed export most of them, because of their food and sports value. The inclusion of native companions (brolgas) on the list of protected native game is perplexing, as they were not favored targets for sportsmen or commercial hunters, nor were they valuable commodities for export.[67] Instead, it is better to see their inclusion in the act as part of a secondary narrative that saw the ASV value birds for their aesthetic appeal. The same impulse that saw the ASV dream of imported songbirds led it to wax lyrical about the virtues of satin bowerbirds and describe kookaburras as possessing "robust, jovial humour."[68]

This quixotic combination of utilitarian self-interest, a desire to protect and develop the food supplies of the colony, and an underlying concern with aesthetics also affected schedule one and the ASV's relationship with colonial hunting culture. Of the animals listed in schedule one, the ASV attempted to acclimatize pheasants, hares, swans, deer, and antelope. Private individuals also tried to introduce all these species (except antelope), setting up a potential conflict over ownership of these animals and different conceptions of how they should be used and managed.[69]

On 30 June 1864, the secretary of the ASV, George Sprigg, wrote to George Watson, the master of the Melbourne Fox Hounds, complaining that the club had hunted and slaughtered a fallow deer, which according to the ASV's interpretation of the 1862 game act remained the ASV's property.[70] This letter was the beginning of a controversy that led to angry letters to the editor, threats and counterthreats of legal action, and the breakdown of cooperation between the Hunt Club and the ASV. Both sides of the dispute believed that deer should be introduced to Victoria and that the deer should eventually be slaughtered. They disagreed over who owned the deer, when it was safe to start culling the deer population, whether deer should be killed primarily for sport or for food, and who should participate in the sport of deer hunting.

Prominent local squatter Andrew Chirnside introduced herds of red deer (*Cervus elaphus*) and fallow deer (*Dama dama*) in the late 1850s.[71] In Tasmania and New South Wales, prominent settlers introduced these same deer species to their properties, both for their aesthetic value and for mounted hunting. The acclimatization societies, both in Britain and in Australia, had a more comprehensive vision for deer acclimatization, envisioning wild populations in isolated areas and looking toward the acclimatization of species of deer and antelope not native to Britain.

The Hunt Club had a much narrower view of deer introduction, viewing them only as quarry for elite sportsmen. The Hunt Club also accused the ASV of violating the act by slaughtering hare and deer for the ASV's annual dinner.[72] Furthermore, it argued that the deer the club killed were not transported to Victoria by the ASV, but were "introduced into the country years ago, in the same way as other animals—horses, sheep, cattle, hares and rabbits—were introduced, before the ill-fashioned word acclimatisation was invented, namely, by private enterprise, without the assistance of Government funds."[73] The club was using the traditional argument that game animals were a form of inviolable private property. The club claimed to own the fallow deer because individual members imported them. It did not claim ownership because of connections to earlier aristocratic traditions of game ownership.

Edward Wilson countered the club's claim to ownership by pointing out that it did not matter who introduced the deer. They remained protected by the act; in effect, he was arguing for public property in animals at the expense of the privileged hunting classes. Defending the ASV's legal action, he asserted that the ASV had little choice but to act. The organization had persecuted poor men for stealing hares and boys for destroying birds. There should be no discrimination according to class or caste. An example needed to be made "of the gentlemen, who should be restrained by other considerations, than in that of his humbler or younger brother, who might probably have erred from ignorance or want of consideration."[74] Legal restraint was necessary to contain the destructive urges of the upper classes when moral sentiment had failed.

As asserted by Wilson, very few of the decision makers within the ASV were sportsmen, although a certain number of subscribers hunted recreationally.[75] Wilson himself frequently swore his disinterest in hunting, shooting, and fishing. When defending the ASV's legal action against the Hunt Club, Wilson wrote:

> I am no sportsman. I never saw a fox hunt, or a stag hunt, or a hare coursed, or a pheasant shot, or a salmon hooked. From a simple wish to see the colony rapidly stocked up with such beautiful things, to add to its attractiveness, to afford scope and variety to the pleasures of these very gentlemen, we are working earnestly and laboriously, and on the whole successfully.
>
> . . . The sports of the field develops the finest race of men physically that can be seen anywhere, but to preserve the *prestige* usually attached to the name of the British sportsman, physical qualities are not all that is necessary.[76]

Wilson was praising hunting culture in theory, but continuing a long-held liberal critique of hunting in practice, namely, that it led to moral degeneration.[77] Furthermore, there was a fear that British hunting traditions were degenerating to selfish farce in the Australian colonies. The ASV and the Hunt Club pursued divergent visions of hunting in Australia, one replicating British aristocratic sporting traditions using familiar animals, the other creating a new heterodox tradition, using animals from all over the empire.

Because of its heterodox views, by early 1864 the ASV was somewhat disillusioned with the 1862 game act and advocated for new legislation. The 1864 Act to Consolidate the Laws for the Protection of Fisheries and Game demonstrates how conflicted the ASV was regarding the nascent hunting cultures of Victoria. Colonial hunters simultaneously wanted to regulate commercial hunting and encourage sports hunting. Examining the reactions helps place the consolidation act in its particular political and social context.

The 1862 game act and collective ownership of game animals proved difficult to enforce. The ASV wrote to the attorney general suggesting scrapping the entire 1862 game act because it failed to prevent "the wanton and useless destruction both of imported and native animals." A new act would give the governor in council the power to create specific closed seasons for individual species.[78] The ASV was not alone in believing that the 1862 legislation was ineffective and in need of revision. In 1863, the *Bendigo Advertiser* pointed out that colonial magistrates inconsistently enforced the act.[79] Furthermore, it argued, in many districts, the act was "nearly a dead letter, turkeys and other game being constantly and openly killed without anyone seeming to care about it."[80] The *Advertiser* and the ASV thought the 1862 game act was a mess in need of extensive revision and enforcement.

In April 1864 the consolidation act was passed; unsurprisingly, it consolidated the fisheries and game laws. The new act's sections dealing with game remained mostly unchanged, except for an amendment suggested by the president of the ASV, giving the governor in council, in consultation with the ASV, the ability to set closed seasons and add species to the protected list.[81] This decision enshrined the ASV's involvement in species protection and dealt them into most of the nineteenth-century Victorian debates regarding fauna conservation.

Differing visions of land use and hunting traditions continued to affect attitudes toward the game laws. In 1862 another land reform bill was passed, setting aside ten million acres for agriculture and allowing settlement after survey. There was agitation for further reform.[82] Agricultural settlement, and with it the culture of agrarianism, shifted northward and westward. Some farmers became actively hostile to the ASV because of its support for the game acts and its ambivalent attitude toward ritualized sports hunting.

The ASV's support for the consolidation act was not based solely on utilitarian resource management. Some members, like the ASV's deeply conservative second president, William Haines, did want to re-create aristocratic hunting in Victoria. At the ASV's second acclimatization dinner, Haines gave a provocative speech that indicated he saw the ASV's role as encouraging sports hunting rather than encouraging agricultural development. He declaimed that "although it might be difficult to acclimatise the English country gentleman, it would not be difficult—provided they could acclimatise the game and the sports of the mother country . . . and so make the bush the permanent home of something more than pauper agriculturalists."[83] Haines thought it was possible to replicate the class structure of England through the importation of British game laws and game animals.

Opponents of the squattocracy were skeptical of the consolidation act for just this reason. Reporting on Haines's comments, the *Gippsland Times* argued that

> for the whole history of our land legislation has been but, that of a contest whether the land should support sheep and game alone, or whether sheep should co-exist with man. We have often warned our readers that there is strong desire on the part of acclimatisers to introduce the game laws of England into this free country, and to that part of their programme we shall not fail to offer the most determined opposition, nor to sound the alarm whenever a further attempt to render more stringent the existing law is about to be made. *The game laws and landlordism have been the bane of agriculture in the old country* [author's italics].[84]

These statements by Haines and by the *Gippsland Times* are indicators of both the ASV's ambivalent relationship with British hunting practices and increasing skepticism of acclimatization among farmers. Agrarian reformers saw game laws as a re-creation of injustices from the old country and abuses of liberty that they were determined would not occur in the colony. The rural districts were by no means universal in their condemnation of the 1864 consolidation act. The *McIvor Times and Rodney Advertiser* celebrated the conviction of a "public poacher" in 1866.[85] It maintained that proper enforcement of the game laws would lead to a recovery in wildfowl numbers that were decimated by "continual shooting" during the gold rushes. The newspaper thought the game laws were just because they preserved public access to game and that violators of the act were "little or no better than a robber" of the public purse.[86]

The ASV never completely embraced acclimatizing the sports of the "English Country Gentleman" and remained somewhat skeptical of sporting practices that inhibited agriculture or commercial hunting. The ASV's attitudes to swivel guns illustrate this skepticism. Swivel guns, or punt guns as they were sometimes known, were large double-barreled shotguns mounted on the front of shallow-bottomed boats (punts) and used for the commercial harvesting of wildfowl from Melbourne's swamps.

Swivel guns were essential for efficiently and cheaply bringing wildfowl to market, a cause dear to the ASV. But they were also anathema to sports hunters and potentially a threat to the long-term viability of commercial duck hunting. Swivel guns first became an issue in 1864, when letters appeared in the newspapers decrying their use and declaring that wildfowl "will soon become a rarity, and a most expensive luxury."[87] Attempts were made to amend the 1864 consolidation act to restrict the use of swivel guns. The ASV was first informed about the swivel-gun problem by the parliamentarian Louis Lawrence Smith. He had become aware of the devastation they caused to wildfowl in the Geelong region and consequently advocated their ban.[88] Faced with the problem, the ASV equivocated. Samuel Bindon, an ASV member and a parliamentarian, supported the ban in Parliament. "If such a wholesale destruction of wild-fowl" were allowed, he argued, "in a very few years the colony would be as destitute of game as Western Port now was of oysters."[89] Bindon was harking back to the destruction, in the late 1850s, of the Western Port oysters, as an object lesson for managing wildfowl resources. Nevertheless, the ASV did not, at this stage, support the swivel-gun ban, and when questioned on the matter, the organization disavowed any interest in banning swivel guns.[90] In this instance, the ASV's interest in commercial hunting and the food supply trumped its interest in resource conservation or sports hunting.

As with the 1862 game act, the ASV discovered that it was tough to achieve convictions under the 1864 consolidation act. In early 1866 the organization attempted to have John Martin convicted of shooting one of their hares at Royal Park.[91] The magistrate threw Martin's case out of court without charge after he successfully argued that he mistook the hare he shot for an unprotected rabbit.[92] Sometimes the ASV would go as far as hiring private investigators to look into violations of the consolidation act but to no avail.[93] As late as 1872, the ASV complained that hares "are shot, hunted and coursed, at all times and seasons, in defiance of the game act, to such an extent that had the hare not have been more prolific in this climate than in England, it would have long since been exterminated."[94] It was the increasing abundance of hares, rabbits, and other introduced species that led to a new round of game laws being passed in 1867 and 1872 as exotic animals increasingly became pests, farmers became more assertive about their property rights, and sports hunters came into conflict with farmers.

TOWARD THE END OF THE 1860S, the ASV began to accept the idea that game animals might be defined as property. It did so for several reasons: first, some animals became so well established that their protected status was unnecessary and even become agricultural pests; second, as the land was divided up through selection, and the agrarian frontier pushed further into the hinterlands of the colony, newly assertive farmers began to claim the right as property owners to destroy the pests that threatened their livelihood; finally, sportsmen began to agitate against the use of swivel guns. These problems created a conflict between the ASV's aim of encouraging the development of yeoman agriculture and its vision of a bushland abundant with game and freely available to all. By 1872, this conflict was resolved in favor of intensive farming and sports hunting.

In 1867, the Act to Protect Game was passed without opposition and with the support of the ASV.[95] It separated game and fisheries protection. Most of its provisions echoed the earlier legislation but with one significant change. It had a clause banning the use of nets to kill game.[96] It was the first act to specifically ban certain means of killing game and enforcing "sportsman-like" behavior at the expense of older traditions developed from poaching.[97] In many ways, this act was a transitional piece of legislation. It bridged the gap between the 1864 law, which enforced closed seasons and public property in game, and the 1873 Act to Further Amend an Act entitled "An Act to Protect Game" and to Repeal the Act No. 438, which protected the property rights of farmers and reduced the protection of hares.

The game legislation of the late 1860s and early 1870s must be seen in the context of a newly assertive and expanding farming class and their conflict with the hunt clubs and the ASV. The expansion of farming into new territories was made possible by the 1865 Grant Land Act that enabled some three million acres to be selected at £1 per acre.[98] The Grant Land Act was still abused by squatters to secure the best land within their former runs in the Western District, forcing selectors east into Gippsland and into the northern plains. Thus there was still considerable and unresolved animosity between squatters, many of whom participated in formal hunt clubs, and selectors, who wanted to protect their crops from pests, both introduced and native.

In 1868 Mr. Dunn, a farmer in Essendon, sued the Melbourne Hunt Club for damages done to his crop when one of its meets ran through his property. The case was heard in the County Court but was then referred to the Supreme Court.[99] Dunn's legal team consisted of George Higinbotham and Archibald Michie, both of whom were prominent lawyers and reform-minded liberal members of Parliament.[100] After calling numerous witnesses, Higinbotham concluded his case by arguing that if the jury did not award sufficient damages to Dunn, it would set a precedent that would allow members of the Hunt Club "to go upon a man's land against his will whenever they pleased."[101] The jury decided to allocate minimal damages to

Dunn, in effect endorsing the Hunt Club's right to ride where they might and while undermining Dunn's property rights.

The failure of Dunn's case led some farmers to form an association dedicated to limiting the rights of hunters to pursue game on private property without the owner's permission. They named their organization 'The Fence, Field and Chattel Preservation League." At the league's initial meeting, one supporter "contended the huntsman had no right at all to go over a farmer's ground without first obtaining his permission. He would advise the farmers to stick together, and remember that in Victoria Jack was as good as his master."[102] Another supporter maintained that "the people that hunt here [in Victoria] have no interest in the soil, and but little sympathy with the cultivator."[103] These sentiments demonstrate not only a determination by farmers to protect their property but also an identifiable class antagonism aimed against the forces of organized hunting.

It took the extension of land selection to Phillip Island in 1868, and competition between introduced hares and selectors, to alienate the ASV and farmers. The new selectors objected to the destruction the hares caused to their crops and vegetables. In August 1869 the ASV sent a delegation to Phillip Island to determine local attitudes to the hares. The port master at Phillip Island was of the opinion that 178 of the 180 settlers on the island were in favor of removal of the game from the island.[104] Mr. Fletcher, the ASV's agent on Phillip Island, dissented, arguing that the settlers could be persuaded that the game needed to be protected. A delegation of farmers from Phillip Island met with the ASV in 1870 to discuss the problem of game on the island, including deer and hares.[105] As a result of its delegation, the ASV decided to remove the deer from the island and to harvest as many hares as possible. The ASV's long-term interest in intensive agriculture began to trump its investment in hares. In fact, by the late 1860s several of the ASV's early importations were becoming liabilities for the organization.

In 1871, representatives of the Victorian horticultural societies met with the premier Charles Gavan Duffy to discuss hare protection. They argued that hares did significant damage to orchards and that it was frustrating that it was illegal to shoot hares, unlike rabbits, despite the damage done to their property.[106] Dr. Black, the president of the ASV, said the game bill before the assembly would give the governor in council the ability to remove animals from the game act as necessary, as opposed to the previous act that only gave the governor in council the ability to add animals to the protected list. Furthermore, he remarked, the "Acclimatisation Society was by no means in favour of protective legislation on behalf of animals that had been fairly established here."[107] Duffy concluded that the government would proceed on the principle "of limiting the protection to an imported animal according as it was needed."[108] In effect, the ASV and the government were acknowledging the primacy of selectors' property rights with regard to game animals on their land, a significant reversal of the organization's original policy toward game animals, which emphasized they should be not only available to all but also protected from wanton destruction.

The ASV also reversed its attitudes toward swivel guns. From 1871 onward it advocated the banning of the guns, thus privileging sports hunting over the commercial duck hunters. The organization argued that swivel guns were not only wasteful, because of the number of birds they injured but did not kill, but also unsportsmanlike and consequently interfered with legitimate recreation. Thus the ZASV now argued that it was "highly desirable that a bill should be introduced . . . [for] preventing the wholesale destruction of game, caused by the wanton use of these guns." Equally, it insisted, the guns had the potential to cause the wholesale extinction of wildfowl in the colony.[109]

The ASV formed its new argument within broader discussions about the wastefulness of swivel guns conducted in the press and Parliament. Objections to the use of the guns for commercial hunting focused on the northern plains of Victoria where selectors were displacing large pastoral properties. The newly established Echuca-to-Melbourne train line made it possible and profitable to bring wildfowl and fish to the urban Melbourne and Bendigo markets. The *Riverine Herald* reported that in 1873 the "murderous swivel gun" was used on Lake Cooper by its "unmanly owner" to send over twelve tons of wild duck to the Melbourne markets.[110] The residents of the Waranga District petitioned Premier Duffy to restrict the use of the guns on the basis that they were unsporting and destroyed wildfowl populations.[111] Specifically, the petitioners thought the guns effectively "deprived [the people] of participation in one of the pleasures which this free country at presents affords" and that they were "fast denuding the country of game; that it is an injustice to the great bulk of the population to allow them."[112] The Echuca and Waranga petitioners were calling on the spirit of the earlier Victorian game laws to argue that "one man has been enabled to unjustly interfere" with the public patrimony. In other words, they were highlighting the need for the public ownership of game, albeit with restricted access, combined with British concepts of sportsmanship.[113] The colonial government promised to ban the guns in the upcoming 1873 Act to Further Amend, an Act intituled [sic] "An Act to Protect Game," and to repeal the Act No. 438. In the end, the government did not include a ban on swivel guns in the 1873 game act; a deputation of commercial hunters convinced it that banning the guns would be detrimental to the metropolitan meat supply.[114]

The 1871, 1872, and 1873 amendments to the 1867 protection of game act did address the concerns of the farmers over hares and hunters trespassing on their land, while still preserving many elements of the 1867 protection of game act. The closed seasons for native game remained at the discretion of the governor in council, and the list of protected imported game remained mostly the same, except for the addition of one clause that allowed landowners or occupiers to destroy hares on their property.[115] This new provision legalized the Hunt Club's practice of coursing for hares. In 1873 a closed season for hares was established that limited the rights of landowners to destroy hares. More importantly the Act to Protect Game restricted the public's right to pursue game on Crown land and made it an offense to drive dogs in pursuit of game across private land without the permission of the owner.[116] In effect, these clauses made game on private land the property of the owners and prevented the hunt clubs from pursuing game at will. These amendments were made with the consent of the ZASV.

In the late 1870s and 1880s, the focus of game legislation in Victoria shifted toward protecting aesthetically pleasing, agriculturally useful, and rare native animals. A new generation of scientists arrived in Victoria, and new institutions formed (for example, the Field Naturalist Club of Victoria and the Department of Agriculture) that, once again, shifted the focus of the game acts. Before this transformation, the distinct elements of Victorian attitudes toward game animals can be summarized thus: public ownership of game, regulated access to game, and closed seasons. Throughout the 1870s and 1880s, these principles were amended to protect farmers' property rights, limit hunting to Crown land, and extend protection to aesthetically pleasing, agriculturally useful, and rare animals.

LOOKING AT THE HISTORY OF game legislation has shown that hunting in Victoria was shaped by British sporting traditions, gold rush assertions of common property in game, concerns about declining native game numbers, and protecting the colonial food supply. These factors were argued over by an ever-shifting nexus of hunting clubs, commercial hunters, farmers, and the ASV. Victorian game legislation hinged on public ownership of game, but it was an ownership that was constrained by legislation for protected species and the introduction of closed seasons. The ASV's positions regarding hunting in Victoria were often contradictory and inconsistent. The ASV believed in the introduction of animals from all over the world suitable for hunting, but once established, these animals should not be private property and should be available for all to hunt. This view transformed over time as some of the ASV's importations became agricultural pests and the organization tacitly accepted private property in introduced game on private land.

The ASV's attitudes toward hunting and the hunting culture in Victoria were shaped simultaneously by reactions against and attempts to emulate British hunting practices. They were further influenced by the long shadow of the Victorian gold rush, directly through concerns about wildlife depleted by the gold rush combined with the hunting traditions established by the diggers, and indirectly through post–gold rush attempts at land reform with the avowed aim of producing yeoman-style agriculture. Like in Canada, India, and Southern Africa, hunting and hunting regulation in Victoria were a manifestation of empire that refracted but did not replicate the hunting practices of the imperial metropolis. Chapters 7 and 8 will look at the reconfiguration, narrowing, and decline of acclimatization in Victoria in the last quarter of the nineteenth century. This process occurred in the face of new scientific theories, sensibilities, and institutions combined with increasing skepticism about acclimatization among secondary actors.

CHAPTER SEVEN

The Decline of Terrestrial Acclimatization

By the late 1860s the ASV was riven with conflicts, suffered severe financial problems, and was ideologically challenged by failed acclimatization attempts including alpacas, antelope, and salmon and other acclimatized animals that became agricultural pests, such as sparrows, hares, and deer. Trout acclimatization was still showing promise, but success was decidedly small scale. Observing these developments, most studies of acclimatization in Victoria have treated it as a phenomenon of the 1850s and 1860s that collapsed under the pressure of failed experiments, shifting aesthetic sensibilities, rising nationalism, lack of resources, and introduced species becoming pests.[1] There is much truth in this, but it is not the whole story. It is better to say that a series of crises forced the ASV to reconsider its function in the 1870s and catalyzed the transfer of control to a new generation of scientists, institutions, and administrators. The cumulative effects of all these factors, combined with shifting aesthetics and science, meant that by the mid-1880s the acclimatization of terrestrial animals ceased; fish acclimatization expanded but outside the ASV's direct control; and the ASV was running a zoological garden and continuing to comment on game laws.

The protection of sparrows created tension within the ASV. Sparrow acclimatization was one of the ASV's success stories, but by the mid-1860s it rapidly became apparent that sparrows were not confining themselves to eating locusts, and by 1865 they were destroying crops and orchards at an alarming rate. The havoc caused by sparrows split the ASV into two opposing factions. Professor McCoy continued to support the agricultural utility of sparrows, citing studies conducted in Philadelphia and by the eighteenth-century French naturalist Buffon.[2] The opposing faction was led by the stockbroker and ASV council member Johnathan Binn Were.[3] The dispute bubbled on in the background for several years, but nothing was done. In 1869, at McCoy's instigation, the ASV dissected 180 sparrows and found that they had consumed "mainly insects not grain."[4] Market gardeners and the Horticultural Society of Victoria brought the issue to a head. Both groups argued that sparrows did more harm than good. Eventually, after much testimony from both sides, a resolution was passed by the ASV that "in deference to the opinions of other public bodies the council would not oppose the removal of sparrows from the game act," and sparrows lost their protected status.[5] This concession was a significant blow to the ideas of acclimatization as a corrective to environmental damage, acclimatization as the yeoman farmer's friend, and the theorizing of the ASV's founding council members. It exacerbated tensions within the ASV. Tensions accelerated to a crisis point when the ASV's secretary, Eugene Lissignol, was accused of embezzling funds and was asked to resign.

ALBERT LE SOUËF WAS VOTED IN as secretary of the ASV over five other candidates in May 1870. Le Souëf was a very well-connected figure. He served as the Usher of the Black Rod in the Victorian Legislative Assembly.[6] His father was the assistant protector of Aboriginals on the Goulburn River.[7] By 1875, Le Souëf had appointed his son William Henry Dudley Le Souëf (known as Dudley) as assistant secretary.[8] It was the start of a multigenerational association between the Le Souëf family and zoological gardens in Australia.[9] Unlike previous secretaries, Le Souëf became a dominant force within the ASV. The balance of power shifted in Le Souëf's favor when a rule change meant that the ASV council and presidency changed every year, the secretary's position remained permanent, and the ASV established a zoological garden.

The first practical plan to create a zoological collection was formulated by Le Souëf in June 1870.[10] He requested the allocation of a proportion of the ASV's annual budget to the creation of a zoological collection to "popularise" the ASV, stating that in many European cities, zoological gardens were among the most popular attractions.[11] The ASV's council accepted the proposal, and a motion was passed to start a collection consisting of a cross-section of Australian fauna and herbivorous animals gathered from across the globe.[12] This idea was consistent with the ASV's long-term goals of increasing public awareness of the value of indigenous fauna and the agricultural value of introduced herbivores (it also avoided the expense of feeding exotic carnivores). The ASV put a call out to its members for native animals and began assembling displays of Australian fauna.[13] By 1872, however, the ASV had changed its name to the Zoological and Acclimatisation Society of Victoria (ZASV) and purchased a pair of lions, one jackal, and a leopard: carnivorous exotic animals unsuitable for release into the wild.[14] Even these actions can best be interpreted as a gambit to increase the ASV's popularity and improve its finances rather than an abandonment of acclimatization. The ZASV's annual reports continued to boast about the progress made in sericulture, raising trout, establishing ostriches, and Angora goat farming as well as the expansion of and enforcement of the game laws.[15]

These new projects lacked the wide-ranging vision that characterized earlier experiments. They also focused on hunting and farm animals maintained on private property, not on establishing new animals in the wild.

Albert Le Souëf established a pheasant-breeding reserve near Gembrook (in the foothills surrounding Melbourne), demonstrating continuing passion for acclimatization but also a shift away from using acclimatization to promote yeoman farmers and toward improving recreational opportunities for colonists. Le Souëf first suggested the necessity of a remote breeding facility away from human habitation in 1870.[16] He thought Gembrook would be a suitable site for raising pheasants because "there is an abundance of insect life and [there are] very few natural enemies" in the area.[17] In 1872, the ASV's council endorsed his plan after its first president, Edward Wilson, donated £200 toward the project. By 1875, the pheasants were breeding well at Gembrook, and their eggs were distributed to ZASV members.[18] The pheasant-breeding facility was expanded, walled, and staffed throughout the 1870s. By 1880, the ZASV noted that the birds were beginning to spread beyond the game reserve but that hunters were destroying them at will.[19] By 1884, the ZASV's council was openly wondering whether it was worth putting much effort into acclimatizing birds given the absence of a gun tax and effective protection of birds.[20] Breeding pheasants and quail at Gembrook continued until 1888, at which point the government resumed ownership of the facility after the breeding population of pheasants died as a result of rabbits and rabbit baits.

Pheasants were the final terrestrial vertebrate that the ZASV attempted to acclimatize. It was not the rabbit baits alone that killed off the pheasants. The rabbit plague of the 1880s further eroded the scientific legitimacy of acclimatization, already tenuous after sparrows, hares, and deer became agricultural pests in the 1860s. On a practical level, by the 1880s, the utility of acclimatized terrestrial vertebrates for pest control was questioned by the ZASV, and emphasis shifted toward protecting agriculturally useful native animals to control pests.

THE 1880S HAVE LONG BEEN SEEN as a critical point of disjunction in the history of environmental management, sensibilities, and sciences in Australia.[21] It was also a period of proposed large-scale irrigation projects and the expansion of agriculture into Gippsland.[22] Simultaneous to these specific local factors, there were large imperial and transnational shifts in attitudes toward national nature, extinction, and animal welfare.[23] In Victoria, a new generation of scientists, operating within the ZASV and the Field Naturalists

Club of Victoria (FNCV), drew on the implications of Darwinism and biogeography, as developed by Alfred Russel Wallace and Alfred Newton, to reinterpret and question acclimatization and transform the game acts.[24] Albert Le Souëf and Dudley Le Souëf, the ornithologist Archibald Campbell, and Baldwin Spencer (Frederick McCoy's successor at the University of Melbourne) were essential facilitators of this process. They operated through existing and new institutions.

The FNCV was one of these new institutions. It was created by Charles French (the newly appointed Victorian government entomologist) in May 1880.[25] It aimed to continue the natural history tradition in Victoria, offering members a forum to share their interests, publish findings, and organize field trips. There was membership overlap between the ZASV and the FNCV, including McCoy, Mueller, Baldwin Spencer, and Dudley Le Souëf.[26] It has also been claimed that the FNCV was actively hostile to Darwinism; later research has demonstrated, however, that some members were active Darwinists.[27] It is better to think of the FNCV as agnostic toward evolution, which helps explain the participation of individuals of such antithetical views as McCoy and Baldwin Spencer.

BALDWIN SPENCER (1860–1927) WAS APPOINTED the first professor of biology at the University of Melbourne in 1887. He was a committed evolutionary biologist, a follower of Milnes Marshall at the University of Manchester and Henry Moseley at the University of Oxford. He was also heavily influenced by Alfred Russel Wallace.[28] Spencer took over McCoy's teaching responsibilities in zoology and botany and held his predecessor's character and science in considerable contempt.[29] He was the first person to teach evolutionary biology at the University of Melbourne and later developed a distinguished career as an anthropologist of the Indigenous people of central Australia. Spencer was an enthusiastic member of the FNCV and the ZASV was dedicated to the preservation of disappearing native fauna.

Shifting attitudes toward native and introduced species expressed by local scientific institutions must be seen in the context of changing conceptualizations of extinction and biogeographic distribution. Scientists had accepted the reality of extinction within geological time since the end of the eighteenth century.[30] By the mid-nineteenth century, Charles Lyell's and later Darwin's work helped conceptualize extinction as a process that occurred in historical time. Concrete examples such as the dodo and the great auk helped focus the question of whether extinction should be seen as inherent in the animal itself or caused by environmental factors.[31]

Alfred Russel Wallace wrote about extinction, biogeography, acclimatization, and invasive species.[32] By examining Wallace's article on acclimatization in the *Encyclopaedia Britannica* and his books *Darwinism* and *Island Life*, it can be seen that Wallace simultaneously reinforced and explained Australian experiences. In 1875, Wallace wrote an article on acclimatization for the ninth edition of the *Encyclopaedia Britannica*. According to Wallace, acclimatization could be defined as "the process of adaptation by which animals and plants are gradually rendered capable of surviving and flourishing in countries remote from their original habitats, or under meteorological conditions different from those which they have usually to endure, and which are at first injurious to them."[33] He maintained that there were three possible acclimatization mechanisms: acclimatization by individual adaptation, acclimatization by variation, and acclimatization by heredity.[34] In most instances, however, when exotic organisms were introduced to a new climate, Wallace thought that animals did not acclimatize—gradually adapt to a different climate—but simply possessed the ability to thrive in greater varieties of climate than previously observed. In fact, Wallace observed that many organisms introduced to a new climate "flourish and increase in it to such an extent as often to exterminate the indigenous inhabitants."[35] He drew his evidence from New Zealand case studies where he observed that rats, goats, and pigs were exterminating native animals. Wallace's subsequent books drew on these early data and subsequently inspired Spencer, the ZASV, and the FNCV to advocate for the protection of native wildlife. In *Island Life* (1881), Wallace wrote about the destruction of the forests of St Helena due to the introduction of goats and invasive flora.[36] Several years later, Wallace attempted to generalize and explain displacement, drawing on Australian and New Zealand examples. He argued that native New Zealand rats had been exterminated by European black rats and that imported honeybees were exterminating native bees in Australia. In *Darwinism*, Wallace declaimed:

> The reason why this kind of struggle goes on is apparent if we consider that the allied species fill nearly the same place in the economy of nature. They require nearly the same kind of food, are exposed to the same enemies and the same dangers. Hence, if one has ever so slight an advantage over the other in procuring food or in avoiding danger, in its rapidity of multiplication or its tenacity of life, it will increase more rapidly, and by that very fact will cause the other to decrease and often to become altogether extinct.[37]

This statement is an early articulation of the competitive exclusion principle in ecology and provides a mechanism for explaining the displacement of native Australian mammals beyond vague references to "primitive animals" and "progress."[38] Wallace also attempted to explain why some species acclimatized easily but others died, by reference to a distinct, yet related, argument, "the species that thrive best and establish themselves permanently are not only very varied among themselves but differ greatly from the native inhabitants. . . . So, in Australia, the rabbit, though totally unlike any native animal, has increased so much that it probably outnumbers in individuals all the native mammals of the country; and in New Zealand the rabbit and the pig have equally multiplied."[39] Thus the absence of a native species similar to an introduced species became a predictor of the success of the introduced species, but the presence of similar exotic and native species could also sometimes predict the decline of native species—a conundrum that remains unsolved in modern ecology.[40] These developing hypotheses built upon complex, reciprocal, and sometimes contradictory relationships between metropolitan theory and colonial examples. This can be seen in the different ways scientists understood the decline of native fauna in New Zealand. In the 1870s and 1880s, Wallace, Baldwin Spencer, and the FNCV used New Zealand case studies to justify displacement theory. Meanwhile, in New Zealand, local scientists were beginning to question the validity of displacement theory.[41]

Rabbits, displacement theory, and scientific theory infuse a paper titled "Animals of Australia" presented by Baldwin Spencer in 1888 to the Ormond College Literary and Debating Society. It purported to be a simple description of the history and distribution of Australian animals, but, in fact, it contained an explicit endorsement of Wallace's biogeographical theories and an implicit attack on McCoy's transcendental biogeographical and acclimatization theories. It used displacement theory as a call to arms to protect "primitive" Australian marsupials, not to endorse their extinction. Spencer's paper divided Australian fauna into two distinct categories, introduced and native, and pointed out that "only 100 years ago there was no such thing as a horse, a sheep, a cow, or a fox, or more important still a rabbit, to be found in the length and breadth of Australia."[42] These animals should, according to Spencer, "be regarded as simply usurpers of the rights and lands of the native animals which they will surely do their best, though perhaps unconsciously, to exterminate."[43] "We are as yet little more than on the threshold of our knowledge of Australian forms of life," the paper concluded, "and yet there is no doubt but that the animals, like the aborigines, are doomed to extinction."[44]

There can be little doubt that Spencer thought that Australian native animals were being displaced by placental mammals, sometimes to the detriment of the country as a whole. In doubt, however, were the mechanisms that Spencer believed caused this displacement. Did he, as did Madden in the 1860s, think that introduced species disrupted the economy of nature, or did he believe, as did Wallace, that isolated and evolutionarily primitive organisms were vulnerable to displacement by similar, introduced mammals? The paper positions him within Wallace's orbit, but with some surprising continuities with McCoy's earlier work. Spencer's paper used Wallace's biogeographic regions and his emphasis on evolutionary isolation to explain worldwide animal distributions. When discussing the Australian biogeographic region, he emphasized the dominance of marsupials and monotremes because of long isolation, and the absence of placental mammals. Like many earlier Australian scientists, he observed that many families of birds were absent from Australia.[45] But unlike previous scientists, he did not advocate introducing birds to fill these gaps or to improve nature in any real way. Drawing on McCoy and Owen's research, he still maintained that marsupials in Australia had adapted and radiated to fill the roles of absent placental mammals, but from a strict materialist evolutionary perspective.[46] Spencer concluded the paper by arguing that some marsupials "become carnivorous; some have been developed into gnawing, rodent-like animals; some have become herbivorous and some insectivorous, and yet notwithstanding these varied developments all have retained their common marsupial characters. In this respect the mammals of Australia are the most remarkable of all now living."[47]

His scientific interest in marsupials, combined with the influence of Wallace and his own personal observations of the spreading rabbit plague, led Spencer to believe that Australian marsupials were vulnerable to extinction as a result of competition with introduced species. They, therefore, needed protection. It was an implicit rejection of acclimatization. As a result, Spencer supported the creation of Wilson's Promontory National Park to protect vulnerable native marsupials.[48] In fact, Spencer's idea that introduced species could not live side-by-side with valued native animals, or fill vacant niches in the economy of nature, damaged the accepted scientific underpinning of acclimatization as practiced in Victoria. It also complicates the neat narrative that increased colonial appreciation of native animals in the late nineteenth century resulted largely from sentiment and rising nationalism.[49]

The FNCV, and in particular one of its most active members, the ornithologist Archibald Campbell, together with the ZASV, pushed for the protection of threatened and useful species within the Victorian game acts.[50] Campbell believed it was "fit and proper" for the FNCV to advise the government regarding the game acts, citing two precedents: the Anglers' Protection Society in Victoria and the entomological societies advising on insect plagues in the United States.[51] Drawing on this instrumentalist view of science, in Campbell's view protection ought to include but extend beyond economically important species and acknowledge that introduced species threatened native species. Campbell compiled a list of birds including those that were protected in each colony, and those birds he felt should be protected in Victoria.[52] He was not interested in protecting sports hunting. In common with the precedents established by the ASV, he wanted to protect hawks and insectivorous birds because of their ability to destroy "harmful" insects and pests. All in all, it was a very ambitious list, including over 200 species in total. It was well beyond the scope of previous attempts to protect birds in Victoria through legislation.

In March 1883, the FNCV sought the ZASV's advice on this list.[53] The ZASV referred Campbell's list to a subcommittee, which met with him in April 1883 and recommended the protection of swans and wild geese from January to July every year and that lyrebirds should be protected from January to August.[54] Additionally, it recommended that every bird "down to the coots" on Campbell's list should be protected from 1 August to 30 January each year.[55] Campbell's list as presented to the ZASV has not survived, but it is reasonable to assume that it was based on the list included in his February report to the FNCV. In May 1885 the FNCV, on the advice of the ZASV, struck some birds from the protected list, including "hawks, bee eaters, crowe [sic] shrike's, (except magpie, at current protected), finches, bowerbirds, wattlebirds, leatherheads and parrots, (except swamp or gound parakeet.)"[56] This list reflected the ZASV's preference for not protecting any animals that could be harmful to agriculture (for example, crows, hawks, and fruit-eating parrots). It also demonstrates the continuing influence the ZASV had in determining and enforcing the Victorian game acts. In 1887, following the FNCV's advice, the ZASV recommended that lyrebirds be entirely protected for five years "to give the birds the chance of increasing."[57] Interestingly, by recommending the protection of ground and swamp parrots, they took the first step toward protecting species (if and only if the species were agriculturally benign) based on the possibility of imminent extinction, rather than their suitability for food or hunting.

The ASV was always aware of and concerned about the threatened extinction of some native species. Before the 1883 amendments to the game act, however, extinction threats were described in terms of direct human interventions such as hunting, forest destruction, and water pollution. After 1883, however, Campbell and the ZASV started to frame the potential extinction of native animals because of their inherent inferiority and vulnerability and incompatibility with introduced fauna. Furthermore, they coupled this with the environmental stress on species caused by habitat destruction. According to Campbell, Australian birds "do not appear, like the birds of Europe or America, to be able to adapt themselves, viz., the alteration of the physical features of the country by the advance of civilization and cultivation. Then there is the havoc made with indigenous forests — their natural resorts."[58] Evolutionary primitiveness robbed Australian birds of the fortitude that would allow them to adapt when humans ravaged the environment. Campbell made this argument by acknowledging the problems caused by introduced animals and the destruction of native animals when he argued that laying poison for rabbits kills many native birds, creating a "hard struggle for existence."[59] A mice plague in the Western District was attributed by Campbell to the systematic destruction of birds of prey. Additionally, he felt that the recent Vermin Board conference should have extended its recommendations for the protection of rabbit-eating quolls, goannas, and feral cats to cover owls and raptors.

The 1880s amendments to the game acts advocated by Campbell, the FNCV, and the ZASV, and the scientific theories that underwrote their recommendations, were ambiguous and ambivalent, framed, as they were, among the emerging rabbit plagues. They did not reflect the replacement of earlier traditions with concern for "national nature" and Darwinian science, but the accommodation of new concepts within old institutions. They further reflect incoherent and contradictory reactions to Australian circumstances and agricultural realities, emerging international scientific traditions, and shifting sentiments. The wildlife management strategy that had previously developed in Victoria was challenged and altered by these factors. The rabbit plague was the final nail in its coffin.

THE ESTABLISHMENT OF RABBITS in Australia in the mid to late nineteenth century is one the most cited and studied problems in Australian environmental history.[60] The relationship between rabbits and the acclimatization movement is still, however, poorly understood. Many early attempts were made to establish rabbits in Victoria; they all failed until Thomas Austin released rabbits at his Victorian Western District property.[61] These rabbits established themselves quickly, spread, and inspired other people to introduce more rabbits. By the 1870s they were becoming a nuisance. By the mid-1880s, they were causing a severe agricultural and environmental disaster across vast swathes of the continent.[62] In the beginning, the ASV approved of the introduction of rabbits. In fact, Austin's early success helped convince some ASV members that acclimatization could be viable in Victoria.[63] The ASV bred "silver grey rabbits" at Royal Park (it is unclear whether these rabbits were ever released into the wild), and facilitated the distribution of rabbits by connecting people who desired rabbits with Austin, who would then supply them from his private stock.[64] The "silver grey rabbits" were, however, the private property of Edward Wilson and not the ASV.[65] This fine distinction later helped the ASV exculpate itself and claim that technically it was never involved in rabbit acclimatization. The repositioning started in 1868, when Thomas Black, the president, wrote a detailed letter on the progress of acclimatization to *Land and Water* (Frank Buckland's newspaper). After lauding the ASV's achievements, Black stated that "the society has never liberated" the rabbit, "fearing that it would prove a nuisance to the farmer."[66] This was a rewriting of the ASV's recent history that had as much to do with reorienting the organization's mission as the farmer's friend following the sparrow and hare problems as it did with the new rabbit problem. Nevertheless, by 1869 the official line had become that the ASV would not assist with the spread of rabbits and would deny it was ever involved in their acclimatization.[67] By 1870, Tasmanian settlers were concerned that the "struggle for existence which is continually going on among races of animals" would result in the displacement of sheep by the rabbit.[68] In 1876, Albert Le Souëf disavowed the introduction of rabbits, arguing that "the introduction of the rabbit was a frightful blunder, but it is one the society is not responsible for."[69]

Despite its constant attempts to distance itself from the rabbit plagues, the ASV was repeatedly blamed for the acclimatization of rabbits. In 1884 the ZASV went before Parliament seeking to become an incorporated entity.[70] Several members of Parliament objected to an incorporated entity with the power to indiscriminately introduce animals "which would be a great nuisance, and cause loss to the agricultural population."[71] Specifically, it was pointed out that the ASV was responsible for introducing Indian mynas and sparrows and encouraging the spread of rabbits. ZASV members and sympathetic parliamentarians denied that it was responsible for introducing sparrows, mynas, or rabbits. This was a complete untruth and the zenith of the organization's attempt to whitewash its history.[72]

The Australian and New Zealand colonies canvassed many solutions to the rabbit problem including bounties, poisoning programs, habitat destruction, and introduced diseases. The historian Thomas Dunlap has argued colonists had no way to explain the spread of rabbits because they relied on common sense, observation, and a vague notion of the balance of nature to explain why some animals failed to acclimatize and others "exploded out of control."[73] Spencer and Campbell's observations show that by the 1880s, some colonists had a sophisticated understanding of the mechanisms that led some animals to become pests and others to fail. A combination of Darwinian natural selection, Wallace's biogeographical theories, and local observations indicated to them that there were emerging patterns among invasive pests. The validity of these observations and the illegitimacy of terrestrial vertebrate acclimatization were reinforced and informed by the rabbit plagues in Australia and New Zealand.

Challenges and responses to displacement theory caused by the rabbit plague in New Zealand affected scientific and policy responses to exploding rabbit populations in Victoria. Rabbits reached plague proportions in New Zealand before Victoria. They had started to become a problem in New Zealand by the late 1860s, and by the mid-1870s they were destroying pasture throughout the South Island.[74] Their spread prompted parliamentary inquiries, some limited attempts at legislative control, and practical attempts at mechanical and biological control. The preferred means of biological control soon become the introduction of mustelids—stoats, weasels, ferrets, and mongooses.[75] Frank Buckland, the founding secretary of SAUK, salmon commissioner, and continuing enthusiast for acclimatization, became the principal proponent of acclimatizing mustelids to New Zealand. Alfred Newton was his primary opponent. He believed that the introduction of mustelids would have a devastating effect on New Zealand's unique avifauna and sought to enlist prominent New Zealand scientists to the cause.[76] One letter Newton wrote to Frederick W. Hutton (government geologist and lecturer at Otago University) was republished in the Otago Witness. It warned that New Zealand's "unique and primitive" birds would be killed by mustelids because they have no "instincts whereby they can protect themselves against such blood-thirsty enemies," and introducing mustelids to control rabbits was akin to if "on the complaint of a few Scottish farmers, it were proposed by a foreigner to send over tigers to stop the depredations of the red deer in the Highlands."[77] These comments are consistent with Newton's established views that the extinction of animals via human action was illegitimate and unnatural.[78] It also fits well with Spencer and Campbell's belief that while displacement theory was well founded, the extermination of primitive organisms by placental mammals could and should be prevented. Unfortunately, Buckland, not Newton, won this debate, and mustelids were acclimatized in New Zealand, to the detriment of native birds.

In Victoria, introducing mongoose to control rabbits was seriously contemplated but rejected. The idea was first suggested by William Bancroft Espeut, a resident of Jamaica and fellow of the Linnaean Society, after he read accounts of Australia's rabbit problem in the British press.[79] He was responsible for introducing the mongoose to Jamaica a decade earlier and claimed they had largely ended Jamaica's rat and snake problem and would "rid Australia of rabbits and perhaps kangaroos and dingos [sic]."[80] Addressing concerns that mongooses might be worse than rabbits, Espeut claimed that any loss of domestic poultry or ducklings caused by mongooses would be more than compensated for by the absence of rabbits. The local press vigorously debated his proposal. Some colonists with experience in India thought mongooses were the key to snake control; others thought their effect on poultry and their potential to become a pest species outweighed any possible benefit.[81] One correspondent mentioned how rabbits, sparrows, thistles, and watercress had become pests and how little the colony had benefited from past introductions, and declared that the whole venture of acclimatization was the province of those who "desire to gratifiy [sic] their sentiment regardless of the laws of Nature."[82] Another correspondent pointed out that in India, where mongooses are native, game birds had "true to the Darwinian theory of animals adapting themselves to the circumstances surrounding them" learned to roost in trees away from mongooses.[83] Releasing mongooses in Australia risked unleashing a "scourge impossible to eradicate" upon ground-nesting introduced poultry and native birds.[84] In 1884, when the ZASV was attempting to become incorporated, parliamentarians raised the possibility the ZASV might follow New Zealand's lead and acclimatize mongooses in Victoria.[85] The ZASV assured Parliament that it never had, and never would, introduce the mongoose to Australia.

The ZASV's, and broader colonial Victorian society's, rejection of mongoose acclimatization to control rabbits is a significant political, scientific, and ideological rejection of previous practices. No longer was Australian nature looked upon as something defective and damaged that could be corrected by introducing animals to fill ancient and emerging gaps in the economy of nature. Rabbit control shifted away from the wildlife management strategies long advocated by the ASV and toward chemical and mechanical controls, introducing diseases to control rabbit numbers and protecting animals already established in Australia and known to prey on rabbits.[86]

AFTER THE RABBIT PLAGUE, even prominent members of the ZASV came to see the acclimatization of terrestrial vertebrates as foolhardy.[87] Dudley Le Souëf wrote in the 1890 proceedings of the Australasian Society for the Advancement of Science that: "The subject of Acclimatisation is a record of great successes and great failures, and I regret that my experience of the subject tells me (and mine is the experience of all interested in this subject) that, as a rule, it would have been better for Australia if the great successes had been failures and the failures successes."[88] He regretted that cashmere goats, alpacas, pheasants, and partridges did not become acclimatized; he also regretted that hares, rabbits, foxes, and sparrows acclimatized; he was, however, pleased that trout acclimatized. If history were fiction, this would provide perfect narrative closure to this chapter and, indeed, the book. Much could be made of the trope of a sinner redeemed. It would be possible to argue that the ASV transformed itself from environmental vandal, with its long-held but now abandoned desire of improving the country with exotic organisms based on dubious science, to running a zoo and protecting endangered species. Unfortunately, reality is rarely that pat. Only terrestrial vertebrate acclimatization declined and disappeared in late-nineteenth-century Victoria. Fish acclimatization became institutionalized.

CHAPTER EIGHT

The Transformation of Fish Acclimatization

> The trout rises quickly, there is a swirl on the surface, and when the ripple clears away he is back in the old spot under the lilies, the grasshopper inside him.
>
> It is rather unfortunate that from one point of view the "tastiest" will not "survive" here. Nothing is better eating than a properly cooked blackfish. The English trout are annihilating them, however.
>
> —DONALD MACDONALD, *Gum Boughs and Wattle Bloom*, 1887

Macdonald here describes his youthful fishing trips as a form of rural idyll—Australia as England reborn, better than the original—Edward Wilson's acclimatization vision from the 1860s. When describing his fishing trips, MacDonald waxed lyrical about native and introduced fauna he witnessed, and it is unclear whether he lamented or celebrated declining blackfish numbers. As historian Tom Griffiths has observed, Macdonald viewed rural Victorian nature and agriculture as a unified whole, where both the people and wildlife were, imported and native, living together in a newly established yet harmonious whole.[1] Fish acclimatization, unlike rabbits and sparrows destroying crops, continued to facilitate this harmonious illusion. Fishing allowed for the combination of imperial recreation and appreciation of local wildlife.

Aesthetics alone, however, cannot explain the continuation and expansion of fish acclimatization during the late nineteenth century. Further factors include a shifting to justifications for fish acclimatization, decentralization, and international developments in fisheries science that saw aquaculture as a panacea for declining fish stocks and scientific disinterest in Australian fish.

New developments in aquaculture and the professional managers who oversaw them were well received in Victoria because it was compatible with the ASV's fundamental precepts and the institutional framework it had been instrumental in developing. While there is no historiographical consensus regarding the development of aquaculture in North America and Europe, several scholars have pointed out features that are salient to the study of acclimatization in Victoria. These features include an ongoing tension between recreational and commercial fishing, a feeling that rivers supporting relatively few species should be stocked with species from elsewhere, and the idea that damaged fisheries could be restored not necessarily with the original species but with similar but hardier species from elsewhere more capable of surviving in degraded environments.[2] Developments in fisheries science reinforced and gave greater credence to what the ASV had always believed: Victorian rivers were barren, damaged, and could be improved. These principles governed fisheries' management in Victoria long after the ASV ceased to be an effective force.

Albert and Dudley Le Souëf expended a lot of time and energy attempting to acclimatize trout on the ZASV's behalf. These attempts were not very successful because of the substandard fish-raising facilities built in Royal Park. To counter this problem, in 1874 the ZASV was given a special supplementary grant of £300 specifically to establish a trout-breeding facility at Wooling. It was designed to create an independent breeding population of trout in Victoria, separate from the Salmon Ponds in Tasmania. Despite raising and distributing some trout, the Wooling facility never really achieved its aim due to uncertain water supply, poor maintenance, and the depredations of cormorants and poachers.[3] By 1880 the ZASV had abandoned attempting to breed trout at Wooling and switched its efforts to encouraging private fish-breeding establishments like Samuel Wilson's at Ercildoun and the regional fish acclimatization societies, and coordinating the release of young trout into Victorian streams.

THE REGIONAL FISH ACCLIMATIZATION societies were offshoots of the ZASV that continued acclimatizing fish in Victorian rivers long after the ZASV itself abandoned acclimatization. Their understanding of and motivations for acclimatizing fish drew inspiration from a diverse group of sources including the ZASV, the United States Fisheries Commission (USFC), and local recreational fishermen. Through a combination of these factors, the regional fish acclimatization societies were able to thrive, gain legitimacy, and establish brown trout and English perch in many Victorian waterways, even as rabbit plagues, shifting aesthetics, and new science discredited the acclimatization of terrestrial vertebrates.

Congress created the USFC in 1870; its original purpose was to investigate declining commercial fisheries and make regulatory recommendations.[4] Its purview quickly expanded to include the establishment of federal fish hatcheries, acclimatizing fish species throughout the United States and supplying, on request, fish ova to recreational and commercial fishermen. Over time the USFC instigated scientific studies of fisheries. It also became a storehouse of aquaculture knowledge and organization worldwide. The USFC provided the regional fish acclimatization societies with expertise in aquaculture and an independent supply of ova, separate from the ZASV's network. Crucially, it also presented aquaculture as a panacea for declining fisheries, thus supplying imperatives for fish acclimatization even as the ideological basis for terrestrial vertebrate acclimatization was decaying.

This American institutional influence coupled with a new angling culture that was emerging in the Australian colonies. It was partially a transplant of British and colonial angling practices to Australia, but it was also the result of new ways of relating to and inhabiting Australian nature and landscapes. During the 1870s and 1880s, a new nationalism emerged in Australia. It manifested in many ways, including the formation of national parks and enthusiasm for bushwalking as an antidote to degeneration within a context of increasing industrialization and urbanization.[5] Although the new nationalism venerated some elements of Australian nature, it also embraced nature transformed by agriculture, irrigated, tame, and recognizably English.[6] Fly-fishing trips seeking out trout in rivers surrounded by Australian bush were a very good way of expressing and participating in this culture. As brown trout became established in Tasmania and New Zealand, a new Australasian fishing culture and literature became established that was not simply transplanted English fly fishing.[7] It involved new behavior in both fish and anglers, but paradoxically celebrated connections to England much more strongly than the early Australasian acclimatizers.[8]

Influenced and supported by the USFC, local anglers formed the Geelong and Western District Fish Acclimatising Society (GWDFAS) in 1875. It had the support of two local members of Parliament, John H. Connor and Robert de Bruce Johnstone, and the head of the local botanical gardens, John Raddenberry.[9] Newspaper accounts and GWDFAS's records illustrate its divergence from the original purposes of the ASV. The *Geelong Advertiser* reported that at the society's first meeting, "it was resolved that an *anglers association* [author's italics] should be formed for the acclimatization of fish in the streams of the Western district."[10] This shift in purpose is further illustrated in the differing formal aims of the GWDFAS and the ASV, which were

1. To introduce all desirable kinds of fish and fish ova into all suitable streams and waters throughout the Western and Wimmera Districts.

2. To take all necessary means of successfully acclimatizing such fish, by the construction of hatching-boxes, breeding-ponds and all other apparatus that may from time to time be needful for the purpose of hatching ova and protecting young fish until fit for removal.
3. To put a stop to netting in proclaimed waters and all unfair fishing; and generally, to do all things that maybe [sic] necessary to the success of the Society and conducive to the interests of anglers.[11]

Unlike the ZASV, GWDFAS had no interest in publishing scientific papers, domesticating new animals, or developing new industries. Like the ZASV, GWDFAS was interested in fisheries protection, but recast its mission as exclusively protecting recreational angling. To achieve the aim of stocking "streams and waters" with "suitable fish," the society immediately began building a fish-raising facility in the Geelong Botanical Gardens.[12] This facility was completed by mid-1875. At its launch, John Connor read extracts from *Domestic Trout—How to Breed and Grow Them*, by Livingstone Stone, an American aquaculture expert and USFC employee. Quoting Stone, Connor declaimed that a great deal of profit could be made from stocking trout at very little expense.[13] The GWDFAS began stocking its new facility with brown trout and English perch directly from the Tasmanian salmon commissioners, sidestepping the ZASV's networks and breeding facilities. The GWDFAS helped establish a fish acclimatization network that built upon part of the (Z)ASV's legacy but was separate from the ZASV itself. It stocked rivers and streams in the Western District with brown trout and English perch for over thirty years. In the late 1890s, it also began experimenting with introducing rainbow trout (*Oncorhynchus mykiss*).

The Ballarat Fish Acclimatisation Society (BFAS) became another central element in the newly decentralized fish acclimatization network. Individuals in the Ballarat District, including the prominent Learmonth family, had taken an interest in fish acclimatization since the late 1850s. Early fish acclimatization attempts included translocating Murray cod into Lake Burrumbeet and golden perch, Yarra herring, and Murray eels into Lake Wendouree.[14] Local luminaries created BFAS in 1870, and like GWDFAS soon established its own fish-breeding establishment and purchased brown trout directly from the Tasmanian salmon commissioners. Unlike the GWDFAS, BFAS demonstrated a continuing interest in acclimatizing native fish.

To develop their private fisheries, BFAS sourced ova and young fish from the USFC. The specific ova they sought demonstrated the continuing influence of the ASV's long-term theoretical conceptualization of acclimatization. Specifically, BFAS used the ASV's assertion that there were affinities between geographically isolated species of fish and that these affinities could predict which species might be successfully acclimatized in Victoria. Albert Günther (an ichthyologist based at the British Museum), it should be remembered, believed that whitefish (*Corregenous clupeiformis*) had an affinity with the Australian grayling. In 1884 the USFC arranged for BFAS to receive one million whitefish ova collected by the United States Acclimatization Society from Lake Michigan. They survived the journey to Sydney, but then died en route to Ballarat and were never established in Australia.

While the regional fish acclimatization societies decentralized and transformed fish acclimatization in Victoria, Sir Samuel Wilson established a private fish hatchery on his Western District property, funded by his own fortune. His hatchery operated separately from the ZASV and the regional fish acclimatization societies. He raised brown trout and lake trout and experimented with hatching and raising Californian salmon (*Oncorhynchus tshawytscha*).[15] Wilson published extensively on fish acclimatization and aquaculture and was cognizant of most of the French, British, and American developments in the field. He read all the relevant works on salmon acclimatization he could find, including Frank Buckland and Livingstone Stone's books and the reports published by the SZA.[16] These reports assured Wilson that Californian salmon would be able to survive and thrive in the high temperatures of Victorian rivers.

Buoyed by optimism, Wilson made two attempts to acclimatize Californian salmon in Victoria. In 1874 he arranged for the Acclimatization Society of San Francisco to send 25,000 salmon ova to Victoria because he had heard that they might survive in much higher temperatures than Atlantic salmon.[17] All 25,000 ova died in transit. In 1877 Wilson tried again, this time with the assistance of the Auckland Acclimatisation Society and Spencer F. Baird from the USFC. Together they shipped 50,000 Californian salmon ova to Victoria acquired from the USFC's McLeod River station in California. Once the ova from his second Californian salmon experiment hatched, Wilson was very careful to select only suitable rivers in which to release the fry. His three principal criteria were suitable temperature, clear connection to the sea, and a lack of industrial pollution. Despite these precautions and some early signs of success, he was unsuccessful in his attempts to establish Californian salmon in Victorian waters. The salmon introduced to New Zealand, however, did much better and were soon established as self-sustaining populations.[18]

Simultaneous with Wilson's attempts to introduce Californian salmon to Victoria, the native Australian grayling population fell considerably. This decline occurred despite the fishery acts and protective interventions from the ZASV, angling clubs, and the fishery inspectors. The Anglers' Society attributed the decline of Australian grayling in the Yarra and the decline of the Yarra fishery more broadly to the river's "gradually being rendered uninhabitable for fish, owing to the wholesale pollution that is being carried on," and to continued inadequate enforcement of the Fisheries Act.[19] In the early 1870s, it took the French consul to Victoria and ichthyologist François Laporte, Comte de Castelnau, over two years to acquire a specimen of the Australian grayling. He claimed, "No specimens, are to my knowledge, now found near Melbourne, and it has become very scarce even in the upper parts of the river" and attributed this decline to the introduction of Murray cod to the Yarra in the late 1850s.[20] In Tasmania, the decline in Australian grayling populations was attributed to predation by cormorants and introduced salmon and trout.[21] In 1878, Samuel Wilson thought that Australian grayling were "threatened with extinction" due to poorly regulated fishing.[22] His proposed solution was the protection of immature fish and rigorously enforced closed seasons, as well as the acclimatization of exotic salmonids to replace grayling. The decline of the grayling reflected traditional ASV concerns, particularly an economically valuable native species being squandered by inappropriate harvesting techniques. However, the response was complicated by new concerns about extinction.

It is worth asking why the potential extinction of Australian grayling in the 1880s by overfishing, predation, and pollution was regarded so differently to the possible extinction of Australian mammals and birds during the same period. Several factors contributed to this divergence. Australian grayling were considered to be taxonomic kin to European salmonids. More broadly, Australian fish were not considered primitive or alien, in stark contrast to Australian marsupials. In fact, superior predatory Australian fish such as Murray cod were often held responsible for impeding the acclimatization of European fish in Victorian rivers. When Australian grayling numbers did start to decline, it was an easy and logical leap, given their supposed taxonomic kinship with salmonids and the availability of European and American salmonid ova, to suppose they could be replaced using the favorite panacea—artificial incubation and the release of salmonids into rivers. Australian scientists were uninterested in studying grayling, leaving recreational fishermen the only group concerned with the species.

Given this level of interest, the 1880s could legitimately be seen as the zenith of fish acclimatization in colonial Victoria, even though terrestrial vertebrate acclimatization was rapidly being discredited. Although the

ZASV shut down its Gisborne trout-breeding facility in 1880, it still enthusiastically endorsed the regional fish acclimatization societies. These societies stepped up their trout acclimatization programs during this period, successfully establishing brown trout and experimenting with American fish varieties.[23] As previously noted, the regional fish acclimatization societies emphasized recreational angling over commercial fisheries but continued to believe that Australian rivers were barren and damaged. They were able to pursue their work effectively because in the 1880s fisheries management matured as a science, and fish acclimatization was widely endorsed as a means of restoring damaged fisheries in America and Europe. This panacea was pursued by a new generation of fisheries scientists who wrote reports, conducted inquiries, and professionally managed fisheries.[24] In Victoria, the professional management of the fisheries began with a parliamentary report called the General Report on the Fisheries of Victoria commissioned in 1887 and authored by William Saville-Kent.

Saville-Kent was part of a new generation of professionally trained fishery scientists emerging from Europe, Britain, and North America in the late nineteenth century. These men were part of a trend toward professional resource management in the Australasian colonies, in the broader empire, and beyond.[25] Saville-Kent received his first Australian appointment on T. H. Huxley's recommendation. Like Huxley, Saville-Kent greatly admired the USFC. This perspective greatly influenced Saville-Kent's first Australian appointment: reviewing fishery and aquaculture practices in Tasmania and implementing the recommendations of the 1882 Royal Commission into the Fisheries of the Colony.[26] Saville-Kent revised Tasmanian fisheries legislation, recommended the establishment of experimental oyster beds, and promoted further investigation into the Tasmanian fisheries. Despite his industry, Saville-Kent left Tasmania in 1887 after falling out with local authorities over the success of salmon acclimatization in Tasmania.

Saville-Kent commenced work on the Fisheries of Victoria parliamentary report in August 1887. He traveled around the state interviewing fishermen, fisheries inspectors, aquaculture experts, and acclimatizers, and by December he had compiled his final report. Saville-Kent's report is one of the few episodes in the history of the Victorian fisheries studied in any depth.[27] His proposals have been well documented: reviving the oyster industry through creating artificial beds, increased protection of freshwater and marine fisheries by incorporating and funding existing angling and acclimatization societies into protection regimes, the artificial cultivation and distribution of native and exotic fish species in Victorian waters, and establishing a deep-water trawling industry.[28] Unfortunately, however, Saville-Kent's recommendations have been treated as if they were the result of the wholesale importation of modern fisheries science from Britain, rather than a continuous interchange between a range of groups: European and American authorities, local scientists and fishermen, and Australian acclimatizers. This interchange was occurring as far back as the first Victorian fisheries legislation and reports in the late 1850s.

The continuities between Saville-Kent's report and the acclimatization and fisheries practices established over the previous decades are apparent in the details of his report and the consultations he conducted with key stakeholders. When discussing developing inland fisheries, Saville-Kent wrote: "At the present time there are numbers of lakes and rivers throughout the country which are absolutely barren," reflecting the ASV's assertions that Victorian waterways were both empty and damaged.[29] The decline of the Victorian fisheries became apparent to Saville-Kent when he visited Sale and the Gippsland lakes, "such decadence of the fisheries being undoubtedly due to overfishing and the destruction of the fish in their breeding grounds," echoing the ASV's long-term concerns with overfishing.[30] Moreover, Saville-Kent's solution to overfishing "artificial propagation and the strict conservation of natural spawning grounds" was part of the same dialogue between imperial and local authority that had been shaping fisheries policy in Victoria for decades.[31] Saville-Kent went so far as to use taxonomic affinity between species to justify acclimatization. Upon being informed of the discovery of the Murray herring (*Nematalosa erebi*), Saville-Kent suggested it could, like its "American congeners," the shad (*Clupea sapidissima*) and alewife (*Clupea vernatis*), be "artificially propagated by millions."[32]

Saville-Kent paid more attention than the ASV to the artificial propagation and translocation of native fish species. The Murray River and its tributaries, in Saville-Kent's opinion, "teem with an immense number and variety of fish, many of which are well worth cultivating."[33] He wanted the regional fish acclimatization societies to breed and distribute Gippsland perch (*Macquaria colonorum*). The Australian grayling was a particular favorite of Saville-Kent's. They should, in his opinion, be stocked into the Hopkins River and the Moorabool River near Ballarat. While in Tasmania, Saville-Kent conducted experiments in artificially breeding Australian grayling and releasing them into rivers. He was skeptical of the ability of Atlantic salmon (*Salmo salar*) to acclimatize in Australian waters but thought the Californian salmon might be possible due to its tolerance of higher temperatures. The translocation of Murray cod might, he believed, be problematic due to its voracious nature. Saville-Kent's recommendations went largely unheeded, except for the establishment of more regional angling societies and some reform concerning measurement as a means of judging fish's eligibility for harvesting based on size rather than weight. His report, however, did legitimatize the fish acclimatization societies and positioned their activities within the scientific mainstream. Science, the regional fish acclimatization societies, and the popularity of angling ensured the continuance of fish acclimatization in Victoria even as the rabbit plague was finally killing terrestrial vertebrate acclimatization.

Two more parliamentary inquiries (1892 and 1909) increased state control over fish acclimatization. The 1892 inquiry recommended more active control over fish acclimatization, but nothing was done. The 1909 inquiry led to the formation of the Fisheries and Game Branch of the Department of Agriculture and recommended extensive aquaculture of both native fish and salmonids. Following this, the Fisheries and Game Branch founded over fifteen hatcheries, often at rural locations and at the request of local angling clubs.[34] These facilities produced a vast amount of brown trout and later rainbow trout ova, establishing trout populations throughout Victoria. The regional fish acclimatization societies maintained their hatcheries but were overwhelmed by the sheer amount of ova the state could produce and distribute. Fish acclimatization and aquaculture became a state prerogative in the early twentieth century.

BY 1900 ALL OF THE ORIGINAL LEADERS of the ASV were dead; many of the second generation had also died or were in their dotage. Many generations of the animals they were responsible for introducing (sparrows, Indian mynas, trout, hare, and deer) had established themselves in rapidly changing landscapes, to be alternately cursed and treasured by colonists, who had just survived a decade of drought and economic turmoil. If the ASV's leaders were able to look back at 1860 from 1900, they would see great transformations in the Victorian landscape, agriculture, science, hunting traditions, and fisheries. In 1900 Victoria was about to become a state of the newly formed Australian nation. The gold rush had long since passed. Land reform had helped some people to create productive farms, but many more had gone bankrupt when faced with poor soils, isolation from markets, and uncertain rainfall.[35] Irrigation was in the process of transforming northern Victoria; in the service of irrigation, rivers were dammed, and the countryside transformed. Despite endless attempts at regulation, fisheries were continuing to decline, and native animals had become rarer. The optimism of the 1880s land boom had been replaced by drought and economic recession, retarding the development of science and limiting the resources available for fish acclimatization.

Despite limited resources, the Fisheries and Game branch was a very active organization. One of its first activities was to establish a trout-breeding facility within the zoological gardens at Royal Park. This facility was jointly funded by the Piscatorial Council of Victoria and the state government.[36] It was operated by Dudley Le Souëf and the ZASV, making it the last acclimatization facility they operated. Between 1908 and 1913, when it was shut down in favor of a state-run facility at Studley Park, the Royal Park facility supplied brown trout and rainbow trout to local rivers.

Rainbow trout were the last exotic species experimented with by acclimatizers in Victoria. The precise details of how and why rainbow trout were introduced to Victoria will never be known in any detail, not only because of the destruction of documents over the last hundred years or so but also because many of the early introductions were conducted by private individuals who left their efforts unrecorded.[37] It is known, however, that rainbow trout were introduced to Victoria from the New Zealand populations that had been established there in the 1880s using broodstock originally supplied by the USFC. The first Victorian rainbow trout broodstock was established in the breeding ponds of the Geelong and Ballarat fish acclimatization societies, which in turn supplied the Royal and Studley Park hatcheries, from whence rainbow trout ova were distributed all over the state.[38] It is not so much the introduction of rainbow trout that is interesting, but the new reasons people gave for opposing their introduction. Fish introductions had been opposed before because of impracticality, climatic unsuitability, taste, or personal preference; never before was an introduction opposed because of its potential effect on native fish stocks.

Donald Macdonald had argued since the 1880s, possibly with regret, that trout were displacing blackfish.[39] Saville-Kent also wrote about the necessity of protecting and artificially breeding native fish in his 1887 report on the Victorian fisheries. These criticisms remained muted, however, until after the establishment of the Fisheries and Game branch in 1909 and the proposed expansion of rainbow trout acclimatization. Some individuals simply thought that rainbow trout acclimatization would be a waste of time due to climatic unsuitability and the species' supposed tendency to migrate to the sea to spawn and never return.[40] Others, Macdonald included, saw rainbow trout acclimatization as futile or unwise because either the rainbow trout would be wiped out by introduced or native perch, or they would eat or otherwise outcompete blackfish.[41] These concerns started a decade-long debate on the cultural and scientific value of native and introduced fish among anglers, naturalists, and the Fisheries and Game branch.

These debates hinged around different responses to displacement theory, a reassessment of the culinary virtues of blackfish, and the reemergence of the tensions between fishing for sport and fishing for food that characterized the hunting debates of the 1860s. By the turn of the century, however, Saville-Kent and Macdonald, among others, were singing its praises.[42] Macdonald kept the debate about blackfish alive, publishing reports by A. H. Moore in Tasmania about the destructive effect of trout on blackfish populations, and colonists' fears that trout "may eventually turn out a pest like the rabbit."[43] Macdonald thought that trout stocking should be limited to mountain streams that did not contain blackfish. He did, however, question whether blackfish could be saved: "The question of the survival of the fittest governing all vegetable, animal and life of the finny tribes presents to us the question. Can we artificially and profitably preserve the blackfish? If we can, is it worth it? Will our sportsmen remain here or go to other States where they have stocked trout for years? I think I am right in saying the brown trout is the best asset, and even without him the days of the blackfish and many others are numbered."[44] Macdonald's column is an almost perfect articulation of both displacement theory with regard to native fish and the cultural acceptance of sports fishing in early-twentieth-century Victoria. Although the decline of blackfish numbers was considered to be scientifically inevitable, Macdonald, the Piscatorial Council, and the Fisheries and Game branch all thought the decline might at least be delayed by enforcing closed seasons and minimum fish sizes. There were also campaigns to prevent stocking trout and perch in new rivers and streams, thus thwarting the appropriation of waterways by trout and elite sports fishermen. Either way, the accusation that trout decimated blackfish populations required a response from the Fisheries and Game branch. The task of responding to these accusations fell to Acting Chief Inspector of Fisheries Frederick Lewis. Whether talking to Macdonald or the FNCV, his defense was the model of consistency. He argued that blackfish numbers were declining in all rivers, whether trout were present or not.[45] This decline was attributed to habitat destruction as waterways were cleared, as well as to amateur anglers who took blackfish "by the sugarbag full" with no regard for closed seasons or size restrictions.[46] Lewis further argued that the "blackfish provides no sport" and that creating a hatchery for blackfish out of "mere sentiment" was a waste of money that could be more efficiently spent on trout acclimatization.[47] For Lewis,

Macdonald, and the Fisheries and Game branch, introduced trout were the best asset local rivers had to offer to sportsmen, whose recreational preferences outweighed scientific worries about exterminating blackfish. The stocking of brown and rainbow trout by the Fisheries and Game branch continued unabated.

EVEN AS IT BECAME apparent that native fish populations were falling, and critiques of both fish and animal acclimatization were being published, Victorian waterways continued to be stocked with brown and rainbow trout, as they are still today. The cultural appeal of recreational angling was too strong, the faith in the panacea of artificial fish stocking too universal, the aesthetic of Australian bush combined with rivers full of European sports fish too appealing to early conservationists to save declining native fish stocks.

The continuance and narrowing of fish acclimatization in late-nineteenth-century Victoria complicates narratives about the veneration of nature, nationalism, and environmental change during the period. When investigations go beyond aesthetics, they reveal that there was no sharp distinction between the utilitarian and sentimental 1860s and a nationalist and Darwinian late nineteenth century. Rather, there was a fusion of traditions expressing pride, regret, and fear for the future. The emerging Australian nation was inextricably bound to introduced species and environmental change to feed and understand itself, while also constrained by and aware of past mistakes. As Donald Macdonald reflected, "In a country like Australia, Nature—working as some think by inscrutable design, others by beneficent chance—had established her own balance before the white man came. Everything we do to upset that balance, which has worked out to perfection through many centuries, merely increases our own difficulties."[48]

WE ALL MUST LIVE WITH these difficulties and the hybrid landscapes that Victorian acclimatizers helped create when seeking to understand, repair, and create a more benign Australian nature using every animal the world had to offer.

Epilogue

"No Sleep Till Hippo"
—Hippo Bill, *The Dollop*

Today, I am not running. I am sitting in a café at the Melbourne Zoo in Royal Park. It feels appropriate to write this epilogue here, where so much of the action in this book occurred. On this bright spring day, listening to animal noises and families chattering, my mind drifts. I contemplate all the acclimatization activities that took place here, the invisible ideas and connections to other continents that made it possible, and the hubris that made repairing and improving Victoria through acclimatization seem like a good idea.

I started this book by contemplating what a casual colonial visitor to the Royal Park acclimatization depot could see in 1861, what an engaged reader of colonial newspapers and attendee of public meetings could understand about acclimatization, and what actions and absences were shaping acclimatization that were beyond the perception of most colonists. Revisiting this conceit but from the perspective of an engaged Australian citizen from 1900 is a worthwhile exercise. In 1900 a casual visitor to Royal Park would see a modern zoological garden complete with animals from all over the world and Australia; while engaged in an elephant ride, she could have observed Assistant Director and acclimatization skeptic Dudley Le Souëf busy at his work. She would not have seen any preparations for releasing animals into the wild or any of the original members of the Acclimatisation Society of Victoria stumbling around the park. Generational change had occurred, and the founding of the ASV and the politics, science, vulnerabilities, and social sensibilities that shaped it were fast fading from living memory. New more nationalistic politics was emerging, along with new science and attitudes toward the natural world. If she were to attempt to read about acclimatization in the newspapers and scientific press, she would find minimum coverage, and what little there was would focus on the damage done by acclimatization and would have discussed trout breeding by local angling societies. By this early stage, the complexities of acclimatization had been forgotten, and acclimatizers were remembered merely as environmental vandals.

When Edward Wilson founded the ASV in 1861, harming the local environment was far from his intent. He and the other founding members had vague ideas of improving and repairing the balance of nature in Victoria by importing animals from all over the world for farmers, hunters, and fishermen. The scientists within the group could not agree on how acclimatization worked but agreed on introducing climatically appropriate animals from all over the world. Their confidence grew as early projects were successful, and they helped develop a global network of acclimatization enthusiasts. After sparrows and hares became pests, this fervor was not matched in broader colonial society. Farmers and hunters, the very groups the ASV was set up to help, resisted acclimatization and asserted their own visions of how nature should be regulated. The ASV's one real success was salmonid acclimatization. It was originally conceived as a means to revive commercial fisheries but later adopted and expanded by angling enthusiasts who operated outside and displaced the ASV's network.

Acclimatization networks in Victoria were complex because there was never a single idea, person, or practice shaping their formation, activities, decline, and transformation. From the very start, acclimatization was a contested and unstable practice that united scientists, farmers, hunters, and fishermen only over the desire to improve Victoria through ecological imperialism and restore colonial damage through neo-ecological imperialism. It was an evolving practice shaped by contingency, luck, scientific theory, imperial connections, and continual experimentation. Amid all this chaos and confusion, Victorian colonists constantly engaged with three sets of interrelated ideas, material realities, and power structures: science, empire, and nature.

To properly understand acclimatization, it has to be taken seriously as a scientific practice embedded in time and place. The theories and programs proposed by McCoy, Bennett, Mueller, and Madden were shaped by and sought to influence environmental change in Victoria, the emerging scientific cultures of the Australian colonies, and contemporary ideas about environmental change and biogeography. They sought to understand why the distribution of animals in Australia was so different, could it be improved, how colonization had damaged the balance of nature in Victoria, and what should be done about it. While divided by scientific theory, all four men believed, with some variation in emphasis, in an acclimatization practice that combined restoring and improving nature with animals from Europe, South

America, Africa, and Asia. Early acclimatization successes reinforced their theoretical positions and made them ever more enthusiastic for acclimatization. The failure of animals to acclimatize and the fact that some plagued did not change their theoretical position.

It was not until a new generation of scientists and officials, such as Baldwin Spencer, emerged and took control of local scientific institutions that the theoretical underpinnings of acclimatization were questioned. They used Darwinism to question the wisdom of acclimatization and to advocate for the protection of "evolutionary primitive" native animals. Of course, new science did not necessarily equal skepticism of acclimatization. In the case of fish acclimatization, the scientific credibility of fish stocking as a cure to declining fisheries increased during the late nineteenth century, leading, when coupled with a craze for fly-fishing, to the entrenchment of salmonid acclimatization.

Acclimatization in Victoria would not have been possible or conceived of without the British Empire. Edward Wilson made use of imperial connections to found the ASV. He was able to use imperial governors to acquire official support for acclimatization and the aid of the British navy for transporting animals. A network of imperial naturalists made the ASV aware of what potentially useful animals existed within and without the empire. They enabled Victorian acclimatizers to dream of transforming and restoring Australian colonies with the animals from Britain, South America, and Africa. Together imperial officers, in the words of Ferdinand von Mueller, attempted to create "another Indian empire in continental Oceania."[1]

Belief in the British Empire should not be confused with uncritical acceptance of British institutions or Britain itself. Australia was to become both a new empire composed of the best elements from all over the world and a better Britain free from antiquated abuses like the restrictive game and fishing laws. Hunting, angling, and aquaculture traditions in Victoria, like in many British colonies, simultaneously drew from British precedents and reacted against them. Acclimatizers, hunters, and fishermen sought to create a natural world that was full of animals from all over the world that, while controlled by regulation, remained the property of all. This continued until it was challenged by acclimatized animals threatening the private property of yeoman farmers and elite hunters claiming ownership of animals. Longing for Britain and imitation of British practices arose, simultaneous with increasing Australian nationalism, amid the late-nineteenth-century anglers of the regional fish acclimatization societies.

On a fundamental level, acclimatization in Victoria called on acclimatizers to examine what they thought of nature in Australia and worldwide, and their role within it. The colony relied for its very existence on the transformation of country for agriculture and mining. It was one large unintentional experiment in the acclimatization of plants, animals, and people. By the time formal acclimatization started in Victoria, the negative and positive consequences of the earlier unintentional experiments were apparent. Victorian acclimatizers thought they understood the virtues and weaknesses of local nature well enough to protect those native animals that they deemed useful and introduce animals from all over the world to fill perceived gaps in the "balance of nature." They only slowly started to appreciate native fauna on an aesthetic level and how acclimatized animals could be a threat. Acclimatization projected a fixed view of nature and its functions onto a colony that was in constant flux. The introduced animals themselves proved the limits of scientific knowledge and refused to stay in their assigned static roles in Victoria, either disappearing into the bush never to be seen again or adapting, changing, and plaguing in new and unpredictable ways.

I have overstayed my welcome at the zoo café. It turns out you can linger too long over a single latte even in hipster Melbourne. As I walk to the exit, I wonder what you, the reader, having trudged through my prose (or skipped directly to the end as I often do) are thinking about, what studying acclimatization in Victoria can teach us about the future, and how to finish this damn book. Ultimately, the dynamics and trajectories of Victorian acclimatization are lessons in humility and unintentional consequences. Acclimatizers proceeded with the best scientific knowledge available at the time and with the support of established institutions. Today, we are proud of our knowledge of ecology and natural systems; acclimatizers felt equally confident that they understood the natural world. Their failures should remind us of our limitations, that scientific knowledge is always provisional, and that there are always blind spots when contemplating ever-shifting nature. What we do today and tomorrow as we adapt to a world transformed by climate change will shape the future irrevocably.

Acknowledgments

Participating in scholarly life is an incredible privilege. I am grateful for the individual and institutional support that has allowed me to write this book. It has been one of the most amazing, excruciating, but ultimately fulfilling parts of my life.

I would like to thank Dr. James Bradley, Professor Andy May, and Dr. Sara Maroske for reading innumerable drafts and encouraging me. Additionally, I appreciate the generosity of Professor Libby Robin and Professor Harriet Ritvo in taking the time to discuss all things environmental history with me. I would also like to thank the professional and helpful team at the University of North Carolina Press, particularly Brandon Proia.

This book would not have been possible without the institutional support and professionalism of the University of Melbourne, the Centre for the Study of the Inland at La Trobe University, the State Library Victoria, the Public Records Office Victoria, the Geelong Heritage Centre, Shepparton Access, and the kind and efficient librarians at the Baillieu Library. I am indebted to the Rudd/Gillard government and the Australian Labor Party for increasing the amount of Australian Postgraduate Award scholarships available and maintaining and expanding an equitable higher education system.

Thank you Alex Chorowicz, Bron Lowe, Alex Dellios, Meighen Katz, Chloe Ward, and André Brett for all the laughter, in-jokes, and support.

Mum and Dad, thanks for encouraging my intellectual interests rather than insisting I become something useful like a panel beater.

Finally and foremost, my profoundest thanks go to Louise Aardvark Minard for loving and supporting a ridiculous human being like me throughout this whole journey.

APPENDIX

Tables

TABLE 1 Native species suitable for export

Species*	Justification	Attempt to export?
Mammals		
Kangaroo	Food, leather, hunting, aesthetics	France, Italy, India, New Zealand, London, Russia, Mauritius, Java
Kangaroo rat (Wallaby)	As above	India, New Zealand, France, Russia
Bandicoot	Food "superior to rabbit"	
Echidna	Food	London
Birds		
Emu	Food, medicinal oil, hardiness	Calcutta, France, Russia, New Zealand, Algeria
Native turkey	Food	
Wonga-Wonga Pigeon	Food "excellent for the table"	India, London, France
Mallee hens	Food, primarily its eggs	London
Quail (various species)	Hunting	Amsterdam
Magpies	Vermin control	France, London, Mauritius, New Caledonia, Calcutta
Kookaburra	Vermin control, song "merry pleasant notes"	Russia, Mauritius, London, France, Calcutta
Black swan	Beauty, food	New Zealand, France, Russia, Mauritius, Ceylon, Java, Copenhagen, Egypt, Japan, Amsterdam, Calcutta, South America (country unspecified), Sicily

(*continued*)

139

TABLE 1 (continued)

Species*	Justification	Attempt to export?
Cape Barren goose	Not specified	London, Java, Mauritius, Paris, Rotterdam
Wild ducks (various species)	Food, ease of domestication	Madras, London, Moscow
Fish		
Murray cod	Food	London, France, New Zealand
Silver perch	Food	
Golden perch	Food	

Source: *Answers Furnished*, Minute Book One, Minute Book Two.
*I have reproduced the common names as used in *Answers Furnished* to preserve how the ASV understood and referred to the species.

TABLE 2 Exotic species suitable for import

Species*	Justification	Attempt to import?
Mammals		
Hog deer	Not given	Imported from Calcutta, released in southern Gippsland—
Manila deer	Not given	Imported from Manila, sold by tender 1868
Axis deer	Not given	Received from Calcutta, given and sold to Sam Wilson in the Western District
Formosa deer	Not given	Received from a private citizen, sold by tender 1868
Rusa deer	Not given	Never acquired
English hare	Not given	Received from Zoological Society of London, extensively liberated
Cape hare	Not given	Never acquired
Snowshoe hare	Not given	Never acquired
Chinchilla	Fur, climatic suitability	Received from South America, bred in captivity, never liberated
Spring hare	Food, climatic suitability—desert	Never acquired

TABLE 2 (continued)

Species*	Justification	Attempt to import?
Gazelle	Food and sport	Received from Alexandria, never able to acquire a breeding pair
Ourebi	Climatic suitability—grassy northern plains	Never acquired
Gemsbok	Food, climatic suitability—desert	Attempt to acquire from the Cape Colony
Oryx	To eat acacias	Never acquired
Eland	Food, climatic suitability—drought-prone areas	Multiple attempts to acquire from the Cape Colony and Natal, none ever shipped
Koodoo	Not given	Attempted to acquire from the Cape Colony
Rocky Mountain bighorn sheep	Climatic suitability—mountainous areas	Never acquired
Cape sheep	Not given	Never acquired
Elk	To eat acacias	Never acquired
Red deer	Not given	Introduced from England, liberated
Roebuck	Not given	Attempted to acquire from the Cape Colony
Rock rabbit	Not given	Never acquired
Birds		
Robin	Vermin control	Shipped from England, liberated
Hedge sparrow	Vermin control	Shipped from China and England, liberated
Indian myna	Vermin control—grasshoppers	Shipped from Calcutta, extensively liberated
Song-thrush	Vermin control—slugs	Shipped from England, liberated
Serpent eater	Vermin control—snakes	Two birds shipped from Cape Colony, never liberated
Crowned pigeon	Food	Never acquired
Himalayan pheasants (multiple species)	Climatic suitability—alpine districts	Shipped from England, liberated

(continued)

TABLE 2 (continued)

Species*	Justification	Attempt to import?
Crested guan	Food	Two birds shipped from England, never liberated
Curassows	Food	Several birds shipped from England, never liberated
Sand grouse of India and Africa	Climatic suitability—northern sandy districts	Acquired from Algeria, liberated
Adjutant of India	Scavenger	Never acquired
Black-necked swan	Not given	Never acquired
Ostrich	Climatic suitability—northern sandy district	Shipped from France and the Cape Colony, farmed in the Western District—not profitable
Fish		
European carp	Pond fish	Acquired from England, liberated
European bullhead	Pond fish	Never acquired
Gourami	Pond fish	Acquired from Mauritius, died in transit
Atlantic salmon	Not given	Acquired from England, liberated
Brown trout	Not given	Acquired from England, liberated
Char	Not given	Acquired by regional fish acclimatization societies
Grayling	Not given	Not acquired
Insects		
Silkworms	Silk making	Acquired from Bombay, experimental colonies established
Cochineal beetles	Dye making	Not acquired

Source: *Answers Furnished*, Minute Book One, Minute Book Two.

  * I have reproduced the common names as used in *Answers Furnished* to preserve how the ASV understood and referred to the species.

TABLE 3 Edgar Layard's list

Species*	Suitability	Use	Availability
Lepus capensis—Cape hare	Thrive in Australia and the high plains of New Zealand		
Lepus saxatilis—scrub hare			
Lepus Rupestris—Smith's Red Rock hare			
Nondescript "river hare"	Thrive in Australia and the high plains of New Zealand		
Elephant	Unsuitable, unlikely to eat gum leaves	Beast of burden, ivory	
Hippopotamus	Insufficient rivers and lakes in Australia	Food	
Hyrax—genus only identified		Food	Easily available; Grey introduced them to New Zealand
Sur larvatus—forest hog			
Phacochoerus africanus—warthog, or marsh hog	Would thrive, but are fierce and untamable	Food	
Equus zebra	Would thrive in Australia	Domestication	Easily available
Equus burchelli (*Equus quagga burchelli*)	Would thrive in Australia	Domestication	Easily available
Equus quagga (*Equus quagga quagga*)	Would thrive in Australia	Domestication	Easily available
Giraffe	Would survive eating Australian acacias; not worth effort	Food	

(*continued*)

TABLE 3 (continued)

Species*	Suitability	Use	Availability
Aigocerus leucophaeus (Hippotragus leucophaeus) — blue antelope			Very scarce
Aigocerus equinus (Hippotragus equinus) — roan antelope			Very scarce
Aigocerus niger (Hippotragus niger) — blackbuck			Very scarce
Kobus ellipsiprymnus — greater waterbuck			
Kobus leche — lesser waterbuck			
Tragelaphus eurycerus — nakon, or bongo			
Oryx capensis (Oryx gazella) — gemsbok, or oryx	Would suit Australian desert		Easily accessed, semidomesticated
Gazella pygarga (Damaliscus pygargus) — bontebok			
Gazella albifrens (Damaliscus pygargus phillipsii) — bontebok			
Gazella euchore (Antidorcas marsupialis) — springbok	Suitable for Australia		Easily accessed
Antelope Melampus (Aepyceros melampus) — impala, pallah			
Redunca eleotragus (Redunca arundinum) — rheebuck	Unsuitable		
Redunca Lalandii (Redunca redunca) — nagor or red rheebuck			Easily captured
Redunca isabellina (Redunca arundinum)			
Redunca capreolus (Pelea capreolus) — rheebuck			Easily accessed
Redunca scoparia (Ourebia ourebi)			

Oreotragus saltatrix (*Oreotragus oreotragu*)—klipspringer	Suitable for Australian and New Zealand mountains	Food, saddlecloths
Tragulus rupestris (*Raphicerus campestris*)—steenbok		Easily accessed
Tragulus rufescens (*Raphicerus campestris*)—vlakte steenbok		Easily accessed
Tragulus melanotus (*Raphicerus melanotis*)—grysbok		Easily accessed
Tragulus pediotragus (*Damaliscus dorcas*)—blesbok		
Cephalophus mergens (modern name unclear)—duiker		Easily accessed
Cephalophus ptook—the dodger		
Cephalophus carula (*Cephalophus monticola*)—bluebuck		
Tragelaphus sylvatica (*Tragelaphus scriptus*)—bushbuck	Suitable for Australia and New Zealand	Easily accessed
Tragelapha angasii—Angas antelope		Easily accessed
Alcelaphus caama (*Alcelaphus buselaphus*)—hartebeest		Hard to catch
Alcelaphus lunata (*Damaliscus lunatus lunatus*)		
Boselaphus oreos (*Taurotragus oryx*)—eland	Not suitable for England, suitable for Australia	Food
Strepsiceros capensis (*Tragelaphus strepsicero*)—kudu	Suitable for Australia	
Catoblepas gnu (*Connochaetes gnu*)—gnu, wildebeest		Easily accessed
Catoblepas taurine (*Connochaetes taurinus*)—kokoon		
Catoblepas gorgon (*Connochaetes taurinus*)—brindled gnu	Suitable for Australia and New Zealand	Food
Bos caffer (*Syncerus caffer*)—buffalo	Not suitable, untamable	

(*continued*)

TABLE 3 (continued)

Species*	Suitability	Use	Availability
Falco serpentarius (*Sagittarius serpentarius*) — secretary bird		Killing snakes	
Francolinus clamator (*Pternistis capensis*) — African pheasant	Suitable for Australia and New Zealand	Game	Easily accessed
Francolinus Nudicollis (*Pternistis afer castaneiventer*) — red-necked pheasant		Game	Easily accessed
Francolinus afer (*Pternistis afer*) — grey-winged partridge		Game	Easily accessed
Francolinus levaillantis (*Scleroptila levaillantii*) — red-winged partridge		Game	Easily accessed
Bustards — species unspecified		Food	
Guinea-fowl — species unspecified		Food	Easily accessed
Migratory quail — species unspecified		Food	
Passer arcuate (*Passer melanurus*) — Cape sparrow		Controlling insects	
Tortoises — species unspecified		Food	Easily accessed
Mugil capensis (*Myxus capensis*) — mullet		Food	

Source: "Cape of Good Hope," *Sydney Morning Herald*, 30 May 1864, 2.

* I have used the scientific names applied by Layard when he used them and have added common names he used and have added current scientific names in brackets where appropriate.

TABLE 4 Animals Edward Blyth offered to the ASV

Species	Description
Mammals	
Axis deer (*Axis axis*)	
Sambar deer (*Rusa unicolor*)	"Should you require more females of this species"
Barasingha (*Rucervus duvauceli*)	"A noble species much like English red deer"
Hog deer (*Hyelaphus porcinus*)	"In any number"
Muntjac (*Muntiacus muntjak*)	
Indian antelope (*Antilope cervicapra*)	
Nilgai (*Boselaphus tragocamelus*)	
Arabian gazelle (*Gazella Arabica*)	
Four-horned antelope (*Tetracerus quadricornis*)	
Bengal hare (*Lepus nigricollis*)	
Bengal fox	"Pretty little animal, innoxious to poultry"
Bengal porcupine (*Hystrix brachyuran*)	"Excellent eating"
Birds	
Wild peafowl (*Pavo cristatus*)	
Junglefowl (*Gallus gallus*)	
Various pheasants and partridges	
Numerous species of wild geese and ducks	"Including most of the European species, save those of arctic habits"
Fish	
Various species of carp	"Well adapted for the Australian rivers, so often reduced to a series of waterholes"

Source: *Argus*, 20 May 1861, p. 5.

Notes

Abbreviations

Ar	The *Argus*
ASNSW	Acclimatisation Society of New South Wales
ASV	Acclimatisation Society of Victoria
Aust	The *Australasian*
Connor Papers	JH Connor Papers, 373/1/1-2, Geelong Heritage Centre Geelong
Field	*Field, the Country Gentleman's Newspaper*
General Correspondence Subject Files	General Correspondence Subject Files, Fisheries and Wildlife Division, VPRS 12011/0001/1, Public Records Office Victoria, Melbourne
GH	*Glasgow Herald*
Letter Book One	Letter Book One 1862, Letter Books Outwards, VPRS 2225/P0000/1, Public Records Office Victoria, Melbourne
Letter Book Two	Letter Book Two 1862–1863, Letter Books Outwards, VPRS 2225/P000/2, Public Records Office Victoria, Melbourne
Letter Book Three	Letter Book Three 1863–1864, Letter Books Outwards, VPRS 2225/P0000/3, Public Records Office, Victoria, Melbourne
Letter Book Four	Letter Book Four 1864–1866, Letter Books Outwards, VPRS 2225/P0000/ 4, Public Records Office Victoria, Melbourne
Letter Book Five	Letter Book Five 1866–1889, Letter Books Outwards, VPRS 2225/P0000/5, Public Records Office Victoria, Melbourne
Minute Book One	Acclimatisation Society of Victoria 1861–1863 Minute Books, VPRS 2223/P0000/1, Public Records Office Victoria, Melbourne
Minute Book Two	Acclimatisation Society of Victoria 1863–1867 Minute Books, VPRS 2223/P0000/3, Public Records Office Victoria, Melbourne

Minute Book Three	Acclimatisation Society of Victoria 1867–1872 Minute Books, VPRS 2223/P0000/4, Public Records Office Victoria, Melbourne
Minute Book Five	Zoological & Acclimatisation Society of Victoria 1880–1884, Minute Books, VPRS 2222/P0000/6, Public Records Office Victoria, Melbourne
News Clippings	*News Clippings* Re Zoological and Acclimatisation Society 01 Jan 1869 to 1910, Records of the Royal Melbourne Zoological Gardens and Related Organisations, VPRS 8850/P00001/81, Public Records Office Victoria, Melbourne
Owen Papers	Richard Owen Papers, Australian Joint Copying Project miscellaneous series, 2278/18, State Library of Victoria, Melbourne
PG	*Portland Guardian and Normanby General Advertiser*
SMH	*Sydney Morning Herald*
Star	*Ballarat Star*
Ti	*The Times of London*
YA	*Yeoman and Australian Acclimatiser*
Z.A.S.V.	Zoological and Acclimatisation Society of Victoria

Introduction

1. Acclimatisation Society of Victoria (hereafter, A.S.V.), *Fifth Annual Report*, 1867, 10.
2. Gillbank, "Animal Acclimatisation," 297–304.
3. For information on imperial networks and networks of science, please see Lester, "Imperial Circuits," 121–41; Gascoigne, "Science and the British Empire," 47–68.
4. Lever, *They Dined on Eland*.
5. Lever, *They Dined on Eland*, 10–28. Other acclimatization societies were formed in Europe including in Berlin, Palermo, Madrid, Moscow, and Switzerland.
6. The ASV was established through £800 of private subscription money. Between 1862 and 1868, it received annual grants of between £2,100 and £3,787. In addition, in this period, annual subscriptions varied between £538 and £1,772. In 1868 there was a funding hiatus from the government. Between 1869 and 1872, the ASV received annual grants of £1,900, and during this period private subscriptions plummeted. After 1872, the ASV renamed itself the Zoological and Acclimatisation Society of Victoria (ZASV) and established a zoological garden. It was funded from this point by a reduced grant and entrance fees. These figures have been taken from the various published annual reports.

7. Multiple attempts were made to secure eland from South Africa, but none were successful: Minute Book Two, 404–7, 633–39. During the first attempt to ship gourami to Victoria, they all died. A second attempt ended the same. Some survived the journey on the third attempt and were placed in the Royal Park pond and the lake at the University of Melbourne, but none were ever seen again. Minute Book One, 133–38, 338–90, 486–89.

8. For studies of botanical acclimatization in Australia, see Maroske, "Science by Correspondence"; Gillbank, "Of Weeds and Other Introduced Species," 69–78. Mueller published several books and pamphlets on plant acclimatization under the auspices of the ASV. That being said, I have found only three occasions when the ASV got directly involved in the exchange of plants. Because of this, acclimatization in Victoria was almost exclusively a zoological practice. By way of contrast, the Queensland Acclimatisation Society concerned itself almost exclusively with plants.

9. Minard, "Assembling Acclimatization," 1–14; Minard, "Zoological Acclimatisation."

10. Boyce, *1835*, 113–43.

11. Robert Kenny has argued that sheep acted as the foot soldiers of colonization in early Victoria: Kenny, *Lamb Enters the Dreaming*, 175–79. By 1842 there were well over a million sheep in Victoria. The idea of sheep as agents of empire was first articulated in Crosby, *Ecological Imperialism*.

12. For a good general account of Aboriginal resistance to colonization in the Port Phillip District, see Broome, *Aboriginal Victorians*, 1–35.

13. Between 1851 and 1861, the Victorian population increased from 77,000 to 540,000. The discovery of gold stimulated the economy, created wealth, and reordered the pastoral economy. There was a shift toward protectionism and land reform. The standard work on this subject remains Serle, *Golden Age*. Several authors have contributed to scholarship exploring the gold rush and environmental degradation, exploring how environmental change left an ambiguous legacy, both shock and disgust at the damage caused and the feeling that environmental resources should be communal and available for all to exploit. McGowan, "Mullock Heaps and Trailing Mounds," 85–103; Frost, "Environmental Impacts," 72–90.

14. Examples of this tradition can be found in literature from all over the world. Examples include Palmer, *Danger of Introducing Noxious Animals and Birds*; Matthams, *Rabbit Pest in Australia*; Thomson, *Naturalisation of Animals & Plants in New Zealand*; and Serventy, "Menace of Acclimatisation."

15. Marshall, "World of Hopkins Sibthorpe," 10.

16. Serventy, "Menace," 189–97; Gillbank, "Acclimatisation Society of Victoria," 255–71; Lever, *They Dined on Eland*; Dunlap, "Remaking the Land," 303–19; Star, "Plants, Birds and Displacement," 5–21; Dunlap, *Nature and the English Diaspora*.

17. Ritvo, "Going Forth and Multiplying," 411–12.

18. Beattie, "Plants, Animals and Environmental Transformation"; Beattie, "Acclimatisation and the 'Europeanisation' of New Zealand," 100–120; Ritvo, "Going Forth and Multiplying," 404–14; Tyrrell, "Acclimatisation and Environmental Renovation," 153–67; Tyrrell, *True Gardens of the Gods*; Mitchell, "Alpacas in Colonial Australia," 55–76; Minard, "Assembling Acclimatization," 1–14; Minard, "Salmonid Acclimatisation in Colonial Victoria," 177–99; Minard, "Zoological Acclimatisation."

19. Ecological imperialism was originally conceived of by Alfred Crosby as a means by which empires were established by the deliberate and accidental importation of old world plants, animals, and microbes to the new world, aiding colonizers and disrupting the social and ecological worlds of colonized peoples. Crosby, *Ecological Imperialism*.

20. Beattie, "Empire of the Rhododendron," 241–61; Beattie, "Plants, Animals and Environmental Transformation," 219–48; Beattie, "'Europeanisation' of New Zealand," 100–120.

21. Cushman, *Guano and the Opening of the Pacific World*, 75–80.

22. I am particularly influenced by McKenzie, *Scandal in the Colonies*, 3; Ballantyne, *Orientalism and Race*, 39; Lambert and Lester, "Imperial Spaces," 13.

23. Lambert and Lester, "Imperial Spaces," 13.

Chapter One

1. Edward Wilson, *Rambles at the Antipodes*, 183.
2. Edward Wilson, *Rambles at the Antipodes*, 183.
3. Serle, "Wilson, Edward (1813–1878)."
4. The existing acclimatization historiography both overplays and misrepresents Wilson's contributions to acclimatization within the British Empire. Too often it attributes actions taken by the ASV and editorial articles written in the *Argus* to be Wilson's initiatives alone. This approach was first articulated in Gillbank's work: Gillbank, "Origins of the Acclimatisation Society of Victoria," 364–72. It is also problematic because it occludes the initiatives of other ASV members and ignores the fact that Wilson was absent overseas for much of the ASV's life and the fact that Wilson did not assert editorial control over the *Argus* after 1856. This problem is compounded when other authors draw upon Gillbank's work and try to imply a connection between Wilson, the *Argus*, concerns about deforestation, and the ASV; see, for example, Bonyhady, *Colonial Earth*, 164.
5. "Principles of Representation," *Ar*, 29 December 1856, 4.
6. The influence of this idea in Victoria and the feeling that pastoralists were squandering the natural patrimony and retarding the development of civilization has been explored within Waterhouse, *Vision Splendid*.
7. "Principles of Representation," 4. Wilson's discomfort with democracy was shared by other liberal reformers who felt that the colony was not ready for "naked democracy" and who were uncomfortable with appeals to class interests made by newly elected "democratic" members of parliament. Serle, *Golden Age*, 277; Goodman, *Gold Seeking*, 73.
8. Gillbank, "Origins of the Acclimatisation Society," 364–70; Gillbank, "Animal Acclimatisation," 298–300. Gillbank has documented the early history of zoological acclimatization in Victoria, looking at the activities of the Philosophical Institute of Victoria's (PIV) Zoological Committee and the short-lived Zoological Society of Victoria (ZSV) in the late 1850s. Key individuals who would later contribute to the ASV include medical practitioner Thomas Black, botanist Ferdinand von Mueller, and paleontologist Professor Frederick McCoy. The ZSV displayed monkeys, birds, goats,

and deer at the Botanical Gardens. Their activities were curtailed because much of the site was swampland unsuitable for free-ranging mammals; furthermore, many of the animals had to remain caged so they would not destroy the Gardens' botanical collections. The government's grant of a larger and more suitable parcel of land in Royal Park to the ASV led to the transfer of the ZSV's animals to the ASV. In 1861 the ZSV was subsumed by the ASV.

9. Mitchell, "Alpacas in Colonial Australia," 65.
10. Moyal, *Bright & Savage Land*, 85.
11. "The Philosophical Institute of Victoria," *Ar*, 9 April 1857, 6.
12. "Philosophical Institute of Victoria," 6.
13. Clements, *Salmon at the Antipodes*, 180.
14. "The Philosophical Institute of Victoria," 5.
15. Humboldt argued that the distribution of plants across world obeyed specific climatic laws. He argued that plant species were distributed across the earth according to latitude, altitude, and the surface temperature of the earth, mimicking the ways plants are distributed on mountains. In Humboldt's opinion, worldwide plant distribution paralleled the distribution of plants on a mountain in predictable ways. Browne, *Secular Ark*, 42–48.
16. Edward Wilson, *Rambles at the Antipodes*, 49–51.
17. Bronstein, *Land Reform and Working-Class*, 62.
18. Philipp, *Great View of Things*, 22.
19. Edward Wilson, *Rambles at the Antipodes*, 68.
20. "Distribution of Animals," *Ti*, 20 October 1858, 6.
21. "Distribution of Animals," *Ti*, 29 August 1860, 11.
22. Minard, "Assembling Acclimatization," 1–12.
23. "Distribution of Animals," *Ti*, 29 August 1860, 11. Geoffroy Saint-Hilaire made these claims while founding the SZA: Lever, *They Dined on Eland*, 3; Osborne, *Nature, the Exotic*, 6–7.
24. "The Alpaca," *Ti*, 17 July 1858, 10.
25. "Quality of Eland Meat," *Ti*, 21 January 1859, 10.
26. Buckland, "On the Acclimatisation of Animals," 21–22.
27. "Quality of Eland Meat," 10.
28. David Mitchell, "Acclimatisation of Animals," 163–70.
29. Mitchell, "Acclimatisation of Animals," 163–70.
30. "The Late Mr Mitchell," *Field*, 7 January 1860, 14.
31. "Society for the Acclimatisation of Animals, Birds, Fishes, Insects and Vegetables," *GH*, 11 August 1860, 3.
32. "Society for the Acclimatisation of Animals, Birds, Fishes, Insects and Vegetables," 3.
33. "Society for the Acclimatisation of Animals, Birds, Fishes, Insects and Vegetables," 3.
34. Buckland, "On the Acclimatisation of Animals," 25–26.
35. Buckland, "On the Acclimatisation of Animals," 29.
36. Buckland, "On the Acclimatisation of Animals," 28.
37. Buckland, "On the Acclimatisation of Animals," 28.

38. "Distribution of Animals," *Ar*, 19 December 1860, 5.
39. "Distribution of Animals," *Ar*, 19 December 1860, 5.
40. "Distribution of Animals," *Ar*, 19 December 1860, 5.
41. "An Acclimatisation Society for Victoria," *Ar*, 26 February 1861, 5.
42. Osborne, *Nature, the Exotic*, 46.
43. A.S.V., *Rules and Objects of the Acclimatisation Society of Victoria*, 22.
44. Gillbank, "Origins of the Acclimatisation Society," 30–65; Dunlap, "Remaking the Land," 305–6; Osborne, "Acclimatising the World," 148–50.
45. A.S.V., *First Annual Report*, 1862, 22.
46. Minute Book One, 54–55.
47. Mason, *Before Disenchantment*, 175.
48. Walton, *Alpaca*, 11.
49. Walton, *Alpaca*, 11.
50. Minute Book One, 11–12.
51. Andrews, "Ledger, Charles (1818 1905)."
52. Minute Book One, 61.
53. Minute Book One, 58–60.
54. "Alpaca," 10.
55. "Alpaca," 10.
56. Minute Book Two, 490–92.
57. Minute Book One, 182–86.
58. Minute Book One, 182–86.
59. Letter Book One, 220.
60. "Acclimatisation," *Field*, 30 August 1862, 210.
61. Minute Book One, 205–10, 252.
62. A.S.V., *First Annual Report*, 1862, 27.
63. Letter Book One, 368.
64. "Acclimatisation," *Ar*, 23 December 1863, 5.
65. Minute Book Two, 566–69, 661–62.
66. "Acclimatisation Society's Dinner," *YA*, 6 July 1864, 531.
67. Minute Book Two, 662–63; Z.A.S.V., *Proceedings and Annual Report*, 1873, 2: 8.
68. Osborne, *Nature, the Exotic*.
69. MacLeod, *Archibald Liversidge*, 2.

Chapter Two

1. Warwick Anderson noted that during the nineteenth century, acclimatization had a different scientific meaning in England than in France. Accordingly, in France acclimatization was defined as investigations into the continuous process of heritable transformation of organisms in response to an alien environment. This definition was built upon the investigations on the effect of climate on biotic distribution by Jean-Baptiste Lamarck and Etienne Geoffroy Saint-Hilaire. Anderson, "Climates of Opinion," 139–40. Michael Osborne investigated the SZA in depth and concluded that it believed in "limited variability of type," where individual animals could change to adapt to climatic circumstances and pass these changes on to their offspring, thus

allowing for the formation of new races (but not species) and allowing acclimatization to take place. He also thought that for the British acclimatizers it was more a matter of relocating animals from one geographically isolated but climatically similar area to another, rather than altering animals' biological adaptations to climate. Following from this, he believed that the ASV followed the British conceptualizations of acclimatization. Osborne "Acclimatising the World," 135-51; Douglas R. Weiner argued that the Imperial Russian Society for the Acclimatisation of Animals and Plants (IRSAAP) followed the French idea of acclimatization. Weiner, "Roots of 'Michurinism,'" 246-51. Paul Star has argued that Darwinian evolution was the primary framework used to conceptualize acclimatization in New Zealand. According to Star, acclimatization in New Zealand was based upon a particular version of evolutionary materialism called displacement theory: the idea that native fauna would inevitably be "displaced" by introduced animals. The theory was based on a rough synthesis of works by Darwin, Joseph Hooker, and Wallace. Star, "Displacement Theory," 4-10. Tom Dunlap conflated the Victoria and New Zealand concepts and argued that Australian acclimatizers used displacement theory to justify acclimatization. Dunlap, "Australian Nature," 32; Dunlap, "Remaking the Land," 161-64.

2. Maroske, "Science by Correspondence," 5.

3. Powell, *Watering the Garden State*, 70; Lucas, Maroske, and Brown-May, "Bringing Science to the Public"; Maroske, "Science by Correspondence," 161.

4. Lucas, "Mueller's Response to Darwinism," 103-30.

5. McCoy resigned from the ZASV's council in 1873. Z.A.S.V., *Proceedings and Annual Report, 1874*, 3:2.

6. Minard, "Assembling Acclimatization," 1-14.

7. McCoy and Owen corresponded on matters as diverse as Australian fossils and Owen's support for McCoy's knighthood. See Owen Papers, 256, 258.

8. Newland, "Dr George Bennett and Sir Richard Owen," 60.

9. Chisholm, "Bennett, George (1804-1893)."

10. Chisholm, "Bennett, George (1804-1893)."

11. A.S.N.S.W., *Third Annual Report, 1864*, 1.

12. Bennett, "Acclimatisation"; A.S.N.S.W., *Third Annual Report, 1864*.

13. Bennett, "Acclimatisation."

14. Australian Medical Pioneers Index, "Henry Ridgewood Madden"; A.S.V., *Third Annual Report, 1864*, 3.

15. Macintyre and Selleck, *Short History of the University of Melbourne*, 5-18.

16. Rupke, *Richard Owen*, 43-52.

17. Rupke, *Richard Owen*, 43-52.

18. Olsen, *Upside Down World*, 129-31.

19. Grove, *Green Imperialism*; Hutton and Connors, *History of the Australian Environment Movement*; Tyrrell, *True Gardens of the Gods*; Bonyhady, *Colonial Earth*; Barton, "Empire Forestry"; Barrow, *Nature's Ghosts*; Beattie, *Empire and Environmental Anxiety*; Cushman, *Guano and the Opening of the Pacific World*.

20. Louis Agassiz was a Swiss-born comparative naturalist and geologist. He immigrated to the United States in 1847, where he was appointed the first professor of zoology and geology at Harvard University's newly formed Lawrence Scientific School.

He believed in the successive, multiple, geographically isolated centers of creation. Edward Forbes was well regarded in his time and lectured in botany at King's College and at the Geological Society. For a detailed explanation of the origins of McCoy's theory and his relationships with the ideas of Owen, Agassiz, and Forbes, please see my article: Minard, "Assembling Acclimatization." For discussion on McCoy's support of Agassiz, please see Butcher, "Frederick McCoy's Anti-Evolutionism," 226-35.

21. A.S.V., *Second Annual Report, 1863*, 36.
22. Osborne, "Acclimatising the World," 149.
23. "Acclimatisation," *SMH*, 15 December 1862, 3.
24. A.S.V., *Second Annual Report, 1863*, 37.
25. Browne, *Secular Ark*, 137.
26. Browne, *Secular Ark*, 162.
27. Rupke, "Richard Owen's Vertebrate Archetype," 231-51.
28. Browne, *Secular Ark*, 5; Minard, "Assembling Acclimatization," 5.
29. This map is not mentioned in the published version of the talk but features heavily in the account of the lecture published immediately after the event in the *Sydney Morning Herald*: "Acclimatisation," *SMH*, December 15, 1862, 3; A.S.V., *First Annual Report, 1862*, 34-45.
30. "Acclimatisation," *SMH*, 15 December 1862, 3.
31. "Acclimatisation," 3.
32. "Acclimatisation," 3. It is unclear what species McCoy was referring to here. It was most likely the common kestrel (*Falco tinnunculus*), but the species was not specifically identified within the text.
33. A.S.V., *First Annual Report,1862*, 38.
34. "Acclimatisation," 3.
35. "Acclimatisation," 3.
36. Bennett, "Acclimatisation"; A.S.N.S.W., *Third Annual Report, 1864*; Bennett, *Gatherings of a Naturalist*.
37. Bennett, "Acclimatisation," 8.
38. Bennett, *Gatherings of a Naturalist*; Bennett, *Wanderings in New South Wales*.
39. Bennett, "Acclimatisation," 13. Bennett's South African river hogs were most likely *Potamochoerus porcus*, or the closely related *Potamochoerus larvatus*. Babirusas are a pig genus from the Indonesian islands comprising four species commonly referred to as "deer pigs." The wild oxen that Bennett refers to are most likely "guar," *Bos gaurus*.
40. Bennett, "Acclimatisation," 13.
41. Bennett, "Acclimatisation," 32. Bennett first became interested in Chinese pheasants during his 1833 visit to Macao and observation of the naturalist, diplomat, and opium trader Thomas Beale. Bennett, *Wanderings in New South Wales*, 49-59.
42. Bennett, "Acclimatisation," 13.
43. Bennett, "Acclimatisation," 13.
44. Bennett, "Acclimatisation," 13.
45. Bennett, "Acclimatisation," 21.
46. Bennett, "Acclimatisation," 23.

47. Bennett, "Acclimatisation," 23.
48. Bennett, *Gatherings of a Naturalist*, 172.
49. Broome, *Aboriginal Victorians*, 99–114.
50. Bennett, *Gatherings of a Naturalist*, 176.
51. Bennett, *Gatherings of a Naturalist*, 177.
52. For excellent discussions of Marsh and Mueller's mutual influences on each other, see Tyrrell, *True Gardens of the Gods*; Tyrrell, "Acclimatisation and Environmental Renovation," 153–67. Humboldt's influence on Mueller is discussed in Jeffries, "Humboldt and Mueller's Argument," 301–10.
53. Bonyhady, *Colonial Earth*, 164–66.
54. Mueller, "Anniversary Address," 7. The Philosophical Institute transformed into the Royal Society of Victoria in 1859 and continues to operate at its original site today.
55. Mueller, "Address of the President," 4.
56. Mueller, "Address of the President," 5.
57. Mueller, "Anniversary Address," 3.
58. Mueller, "Anniversary Address," 3.
59. Jeffries, "Humboldt and Mueller's Argument," 301–10.
60. Mueller, "Forest Culture," 3.
61. Mueller, "Forest Culture," 15.
62. Mueller, "Forest Culture," 15.
63. Mueller, "Application of Phytology," 19.
64. Powell, *Watering the Garden State*, 69.
65. "Monday, October 16, 1865," *Ar*, 16 October 1865, 4.
66. For a broader discussion of colonists' awareness of the environmental damage caused by the gold rush, please see Frost, "Environmental Impacts," 72–90.
67. Tyrrell, *True Gardens of the Gods*, 24–30.
68. Judd, *Common Lands, Common People*, 139.
69. Judd, *Untilled Garden*, 209.
70. Marsh, *Man and Nature*, 103.
71. Frame, *Evolution in the Antipodes*, 102–3.
72. Minute Book Two, 461–66, 683–84.
73. "Acclimatisation as a Means of Restoring the Balance of Nature: A Paper Delivered by Dr Madden," *YA*, 24 August 1864, 762. Darwinian theory offered a viable explanation for the need for acclimatization, but it was largely ignored by the ASV. In fact, "Acclimatisation as a Means of Restoring the Balance of Nature" was the only instance in which any figure within the ASV used an evolutionary framework to justify acclimatization.
74. "Acclimatisation as a Means," 762.
75. "Acclimatisation as a Means," 762.
76. "Acclimatisation Society Conversazione," *Ar*, 18 August 1864, 5.
77. "Acclimatisation Society Conversazione," 5.
78. Mickle, "Lyall, William (1821–1888)."
79. Minute Book One, 114.
80. A.S.V., *Second Annual Report*, 1863, 9.

81. Minute Book One, 122–24.
82. A.S.V., *Answers Furnished*, 14–15.
83. Minard, "Assembling Acclimatization," 8.
84. "Local Topics: Why Is Australia Odd (Continued)," *Aust*, 3 December 1870, 7.
85. "Local Topics," 7. Doug McCann identified McCoy's pseudonym in his article McCann, "Naturalist Tradition," 209–14.
86. McCann, "Naturalist Tradition," 209–14.
87. Minard, "Assembling Acclimatization," 10–14.
88. A.S.V., 27.
89. Minute Book Two, 651–52.
90. A.S.V., *Fourth Annual Report, 1865*, 7.
91. A.S.V., *Fourth Annual Report, 1865*, 7.
92. Minute Book Two, 698–700.
93. A.S.V., *Fifth Annual Report, 1867*, 10.
94. This species is now known as *Prunella modularis*.
95. A.S.V., *Answers Furnished*, 16.
96. "Acclimatisation as a Means," 762.
97. "Acclimatisation," 3.
98. A.S.V., *Answers Furnished*, 16.
99. Bennett, "Acclimatisation," 23.
100. Bennett, "Acclimatisation," 23–24. The French sparrow report was republished in English within the *Times* and as "Destruction of Small Birds by Farmers, Gardeners and Others," in *The Zoologist: A Popular Monthly Magazine of Natural History*, 7729–40.
101. "Deputations. The Gardeners V. The Sparrows and the Hares," *Ar*, 28 August 1871, 6.
102. "Saturday, December 5, 1863," *Ar*, 5 December 1863, 4.
103. Anderson, "Climates of Opinion," 135–67; Osborne, "Acclimatising the World," 135–51; Star, "Displacement Theory," 5–21.

Chapter Three

1. Minute Book Two, 625–26, 691–92.
2. Grove, *Ecology, Climate and Empire*, 37–75; Cheke and Hume, *Lost Land of the Dodo*; Stockland, "Policing the Oeconomy of Nature," 207–31.
3. Minute Book One, 83–84, 380–90, 404–7; Minute Book Two, 478–80, 633–34, 717–19.
4. Lever, *They Dined on Eland*; Gillbank, "Origins of the Acclimatisation Society"; Rolls, *They All Ran Wild*.
5. Lever, *They Dined on Eland*, 74–77.
6. Minute Book Two, 437–40.
7. A.S.V., *Answers Furnished*, 1.
8. A.S.V., *Answers Furnished*, 1.
9. Minute Book Two, 464–66.
10. A.S.V., *Answers Furnished*, 8.
11. Buckland, "On the Acclimatisation of Animals," 25; Bennett, "Acclimatisation," 20.

12. Olsen, *Upside Down World*, 136–41.

13. A.S.V., *Answers Furnished*, 7.

14. Buckland, "On the Acclimatisation of Animals," 20.

15. A.S.V., *Answers Furnished*, 8.

16. A.S.V., *Answers Furnished*, 14–15.

17. Buckland, "On the Acclimatisation of Animals," 21–22.

18. Beinart, *Rise of Conservation in South Africa*, 28–99.

19. Mennell, *Dictionary of Australasian Biography*, 198–201.

20. Brennan, "Imperial Game."

21. Minute Book Two, 437–40, 668–70.

22. Minute Book One, 151–54.

23. Mennell, *Dictionary of Australasian Biography*, 270.

24. Layard published his notes on the ornithology of Ceylon in multiple parts: Layard, "XVIII—Notes on the Ornithology of Ceylon," 165–76; Layard, "X.—Notes on the Ornithology of Ceylon," 97–107; Layard, "XLI.—Notes on the Ornithology of Ceylon," 446–53; Layard, "XXI—Notes on the Ornithology of Ceylon," 212–18; Layard, "V.—Notes on the Ornithology of Ceylon," 57–64.

25. Layard, "XIII.—Notes on the Ornithology of Ceylon," 123–31; Layard, "XXV.—Notes on the Ornithology of Ceylon," 264–72; Layard, "XXIV.—Notes on the Ornithology of Ceylon," 257–64.

26. Mennell, *Dictionary of Australasian Biography*, 270.

27. Letter Book One, 463.

28. Layard's response to the survey were published in full within the Australian press: "Cape of Good Hope," *SMH*, 30 May 1864, 2. Layard's survey was available to the ASV only after it had completed its own version of the questionnaire, so the ASV was not influenced directly by it.

29. For African scientific traditions, please see Rookmaaker, *Zoological Exploration of Southern Africa*; Rookmaaker, "Zoological Contributions of Andrew Smith (1797–1872)." For discussions of African hunting traditions, please see MacKenzie, *Empire of Nature*; Carruthers, "Changing Perspectives on Wildlife in Southern Africa," 183–99; Van Sittert, "Bringing in the Wild," 269–91; Couzens, "Only Half a Penguin a Day," 207–35; Gess and Swart, "Stag of the Eastern Cape," 48–76; Roche, "Fighting Their Battles O'er Again," 86–108.

30. Skead, *Historical Mammal Incidence in the Cape Province*, 14–28.

31. These influences are displayed in Layard, *Catalogue of the Specimens in the Collection of the South African Museum. Part 1. The Mammalia*, 65–81.

32. Couzens, "Half a Penguin a Day," 210–15; Beinart, *Conservation in South Africa*, 196–234.

33. "Cape of Good Hope," 2.

34. "Cape of Good Hope," 2.

35. Van Sittert, "Bringing in the Wild," 275–78; Beinart, *Conservation in South Africa*, 196–234; Roche, "Fighting Their Battles O'er Again," 90–93.

36. Minute Book Two, 490–92, 635–37, 738, 739; Minute Book One, 145–49; Minute Book Three, 831.

37. Samuel Wilson, *Report on the Ostriches*.

38. "Cape of Good Hope," 2.

39. Minute Book Two, 408–11, 805–7; Minute Book Three, 846–47.

40. Minute Book One, 133–38; Minute Book Two, 496–98; Minute Book Three, 954–55.

41. Minute Book One, 156; "Melbourne," *SMH*, April 11, 1862, 5.

42. Zoological Society of London, *Annual Report, 1861*. This expedition was led by J. Benstead and had over £500 to spend. He was able to acquire single specimens of *Tragelaphus strepsicero, Damaliscus dorcas, Pelea capreolus, Raphicerus campestris, Raphicerus melanotis, Hippotragus leucophaeus,* and *Equus quagga burchelli.*

43. Minute Book One, 403–7, 633–39.

44. Discussed in A.S.V., *Third Annual Report, 1864*.

45. Rideout, "'Handsome Gifts' to a Young Society," 37–43; Hoage and Deiss, *New Worlds, New Animals.*

46. "Death of Mr. H. E. Watts," *Ar*, 9 November 1904, 7.

47. Jones, and Kenny, *Australia's Muslim Cameleers*, 39–42.

48. Butler and Hume, *Catalogue of the Birds of Sind, Cutch, Káthiáwár, North Gujarát, and Mount Aboo*; Pittie, *Birds in Books*, 161–62.

49. Mennell, *Dictionary of Australasian Biography*, 270.

50. Ito, *London Zoo and the Victorians*, 159–60.

51. "Acclimatisation," *Ar*, 20 May 1861, 5.

52. "Acclimatisation," 5.

53. Examples of the usefulness of Blyth's position for assembling relevant acclimatization knowledge can be seen in his publications, for example, Blyth, *Report on the Mammalia and More Remarkable Species of Birds Inhabiting Ceylon*, 155–67; Blyth, *Catalogue of the Mammalia in the Museum Asiatic Society*. For Blyth's involvement in the live animal trade, see Brandon-Jones, "Edward Blyth, Charles Darwin, and the Animal Trade," 145–78; Ito, *London Zoo and the Victorians*, 144–54.

54. Letter Book One, 184, 464–65.

55. A.S.V., *Third Annual Report, 1864*, 43.

56. A.S.V., *Third Annual Report, 1864*, 44.

57. A.S.V., *Third Annual Report, 1864*, 37.

58. Letter Book One, 128, 211.

59. Letter Book One, 128, 211.

60. Letter Book One, 128, 211; Letter Book Two, 289.

61. Minute Book One, 436–36; Minute Book Two, 694–96.

62. Letter Book One, 213–14.

63. Letter Book One, 330–31.

64. Letter Book Two, 209–10.

65. Letter Book Two, 371–72, 684; Letter Book Three, 709; Letter Book Four, 739.

66. Letter Book Two, 371–72.

67. A.S.V., *Fourth Annual Report, 1865*, 5.

68. Hall and Gill "Management of Wild Deer in Australia," 837–39.

69. The reputation of Indian mynas as insect destroyers and their introduction to Mauritius are discussed in Stockland, "Policing the Oeconomy of Nature," 207–10.

70. Kate Grarock et al., "Are Invasive Species Drivers of Native Species Decline," 106–14.

71. Ito, *London Zoo and the Victorians*, 150–55.

72. A.S.V., *Third Annual Report, 1864*, 9.

73. Pheasants were released in the Dandenong ranges, at Phillip Island, and at Mount Macedon. They disappeared without a trace. Minute Book Three, 833, 837, 921–22.

74. Star, "From Acclimatisation to Preservation," 45–48; Dunlap, *Nature and the English Diaspora*, 54–56.

Chapter Four

1. Bonyhady, *Colonial Earth*; Tyrrell, *True Gardens of the Gods*; Frost, "Did They Really Hate Trees?," 72–90.
2. Powell, *Watering the Garden State*, 49; Frost, "Environmental Impacts," 72.
3. Bowen, "Power of Money," 85.
4. "The Murray River Fishing Company," *Ar*, 29 July 1862, 3.
5. Leslie, "Moira Lake."
6. Leslie, "Moira Lake."
7. Bowen, "Power of Money," 85.
8. Bowen, 118.
9. Minute Book Two, 649–50.
10. Presland, *Place for a Village*, 183–85.
11. Bennett, *Fish Markets of Melbourne*, 5.
12. A.S.V., *Third Annual Report, 1864*, 60–64.
13. A.S.V., *Third Annual Report, 1864*, 60–64.
14. A.S.V., *Third Annual Report, 1864*, 62.
15. Cushing, *Provident Sea*, 116.
16. A.S.V., *Third Annual Report, 1864*, 62.
17. "The Institution of a Society of Anglers," *Ar*, 2 October 1862, 7.
18. "Institution of a Society of Anglers," 7.
19. Bennett, *Fish Markets of Melbourne*, 8.
20. "Oysters," *PG*, 31 March 1856, 3.
21. Kurlansky, *Big Oyster*, 54.
22. Rowland, "History and Fishery of the Murray Cod," 210–13.
23. Act for the Preservation of Fish in the Lakes and Rivers of the Colony of Victoria, 1859; Act for the Regulation of the Oyster Fisheries in Victoria, 1859; Act for the Protection of the Fisheries of Victoria, 1859.
24. "The Norfolk Islanders," *Star*, 26 April 1862, 1; "The Fisheries of Victoria," *Courier*, October 6, 1862, 3.
25. "Fisheries of Victoria," 82.
26. "Fisheries of Victoria," 83.
27. Minute Book Two, 516.
28. "The Fishery Committee," *Ar*, 27 May 1863, 3.
29. "The Distribution of Animals," *Ti*, 20 October 1858, 6.
30. A.S.V., *Answers Furnished*, 9.
31. McCoy, "Zoology and Palaeontology," 183.

32. Humphries and Walker, "Ecology of Australian Freshwater Fishes," 4.
33. Humphries and Walker, "Ecology of Australian Freshwater Fishes," 7-8.
34. Rowland, *Overview*, 39.
35. Rowland, *Overview*, 41.
36. A.S.V., *Answers Furnished*, 9.
37. Minute Book Two, 450.
38. Leslie, "Moira Lake."
39. Leslie, "Moira Lake."
40. "Lake Moira Fishermen," *Star*, 5 July 1864, 3.
41. Leslie, "Moira Lake."
42. Leslie, "Moira Lake."
43. Minute Book Two, 453.
44. "Country News," *Ar*, 10 July 1863, 6.
45. "Country News," 6.
46. Letter Book Four, 476, 486.
47. Minute Book Two, 668-70.
48. Rowland, "History and Fishery of the Murray Cod," 207.
49. Minute Book Two, 673-74; Rowland, "History and Fishery of the Murray Cod," 207.
50. Rolls, *More a New Planet than a New Continent*.
51. Fish families not native to Australian rivers include Centrarchidae, Characidae, Chiclida, Cobitidae, Cyprinidae, Cyprinodontidae, Percidae, Poeciliidae, and the Salmonidae. Humphries and Walker, "Australian Freshwater Fishes," 8.
52. Saunders, *Discovery of Australian Fishes*, 36-39, 86-98.
53. Saunders, *Discovery of Australian Fishes*, 36-39, 86-98.
54. Cuvier acknowledged thirteen species within the Salmo genus alone. Günther argued for sixty-one different Salmo species. Current science acknowledges forty-one species of Salmo. Cuvier and Griffith, *Class Pisces Arranged by Baron Cuvier*, 186-88; Günther, *Catalogue of the Physostomi*, 6: 95-255; University of British Columbia Fisheries Centre, "Fishbase."
55. Günther, *Catalogue of the Physostomi*, v.
56. We now know these species are not closely related to salmonids and reside within the families Galaxiidae and Retropinnidae.
57. "Angling in Victoria," *Aust*, 3 December 1864, 9.
58. "The Grayling in Australia," *Mercury*, 16 April 1866, 3.
59. Cuvier and Valenciennes, *Histoire Naturelle Des Poissons*, 18: 346-50.
60. Cuvier and Valenciennes, *Histoire Naturelle Des Poissons*, 18: 346-50.
61. "Australian Fresh-Water Fishes," *YA*, 10 October 1863, 25.
62. *Encyclopaedia Britannica*, 9th edition s.v. "Ichthyology."
63. *Encyclopaedia Britannica*, 9th edition s.v. "Ichthyology."
64. *Encyclopaedia Britannica*, 9th edition s.v. "Ichthyology."
65. "Grayling in Australia," 3.
66. "Grayling in Australia," 3.
67. *Command Papers, Report into the Salmon Fisheries (England and Wales)*, xxii-xxxvi.

68. MacLeod, "Government and Resource Conservation," 166–68.

69. Minute Book One, 166–69.

70. Minute Book One, 166–69.

71. "The Fisheries of the Colony," *Ar*, 11 February 11 1862, 5.

72. Minute Book One, 169–73.

73. Act to Amend and Consolidate the Laws for the Protection of the Fisheries of Victoria, 1862, 261–63.

74. Act to Amend and Consolidate the Laws for the Protection of the Fisheries of Victoria, 1862, 261–63.

75. "Fisheries of Victoria," 84; Act to Amend the Laws Relating to Fisheries of Salmon in England, 1861, 8.

76. Act to Amend and Consolidate the Laws for the Protection of the Fisheries of Victoria, 1862, 221–23.

77. Parliament of Victoria, *Votes and Proceedings of the Legislative Assembly 1862–1863, Session One*, 1: 363.

78. Minute Book Two, 419–22.

79. "The Acclimatisation Society," *Argus*, August 20, 1863, 7.

80. "Acclimatisation Society," 7.

81. "Acclimatisation Society," 7.

82. Minute Book Two, 429.

83. Minute Book Two, 434.

84. Minute Book Two, 434.

85. Minute Book Two, 649–50.

86. "The Australian Pilchards," *Ar*, 7 September 1865, 6.

87. "Australian Pilchards," 6.

88. "Australian Pilchards," 6.

89. "Australian Pilchards," 6.

90. Minute Book Two, 749–51.

91. Minute Book Two, 732–34; Minute Book Three, 773–74, 992–97.

92. Minute Book Three, 923.

93. Tyrrell, "Acclimatisation and Environmental Renovation," 153–67; Frost, "Environmental Impacts," 72–90.

Chapter Five

1. The early accounts include Nichols, *Acclimatisation of the Salmonidae*; Senior, *Travel and Trout in the Antipodes*; Samuel Wilson, *Salmon at the Antipodes*. These early accounts were then incorporated into twentieth-century popular histories that parroted the 1880s accounts, unaware of, or unconcerned by, the extent to which trout angling was overemphasized at the expense of commercial salmon aquaculture. Clements provides a largely narrative account of salmonid introduction in Australia with very little examination of the motivations for early attempts to acclimatize salmonids beyond vague references to the emergence of how an "aristocracy of free settlers was beginning to emerge with time and money to think of more pleasurable pursuits akin to the rural sports of Britain." Clements, *Salmon at the Antipodes*, 39. Walker

and Dunn's accounts emphasize alienation from the local landscape and angling as the predominant motivations behind salmonid acclimatization. Walker, *Origins of the Tasmanian Trout*, 11; Dunn, *Angling*, 82. Academic accounts of aquaculture in Australia have drawn from the popular histories and early accounts, rolled it into broader accounts of acclimatization, and attributed motivation to homesickness, enthusiasm for British recreational practices, and alienation from the local landscape. Dunlap did not directly address fish acclimatization, but he did reinforce the idea that acclimatization was motivated by alienation, sentiment, and the desire to reproduce British sporting practices in Australia. Dunlap, "Australian Nature," 28–30; Dunlap, "Acclimatisation Movement," 308–12. The Tasmanian sociologist Adrian Franklin has explicitly examined trout acclimatization in Tasmania, relying too heavily on popular accounts and overemphasizing angling as a motivation for acclimatization. He argued that trout aquaculture was a result of the Britainization of Australia, predicated on the exploitation and loathing of native Australian animals, and the spiritual and Romantic associations of trout fishing. Franklin, "Australian Hunting and Angling Sports," 45; Franklin, "Performing Acclimatisation," 23–25.

2. Key works on aquaculture in North America include Halverson, *Entirely Synthetic Fish*; Taylor, *Making Salmon*; Towle, "Authored Ecosystems."

3. Tyrrell, "Acclimatisation and Environmental Renovation," 157.

4. Towle, "Authored Ecosystems," 59–61.

5. Atlantic salmon and brown trout have been chosen as the main exemplar of the ASV's aquaculture program because of the persistence with which their acclimatization was pursued, the considerable sums of money expended on salmon acclimatization, and the plethora of published material surviving about the attempts. It was easier to import salmonids than other families of fish because salmonid ova could be easily extracted and transported. Other families of fish were imported as mature specimens, reducing the numbers that could be imported and reducing the chance of them surviving the trip to Australia alive.

6. Nash, *History of Aquaculture*, 54.

7. Nash, *History of Aquaculture*, 54.

8. Nash, *History of Aquaculture*, 56.

9. Levasseur and Kinsey, "Second Empire Legacy," 253–68; Kinsey, "Seeding the Water," 535–38.

10. Dunn, *Angling*, 85.

11. Ashworth and Ashworth, *Treatise on the Propagation of Salmon and Other Fish*; Ashworth, *Essay on the Practical Cultivation of a Salmon Fishery*.

12. Wilkins, *Ponds, Passes, and Parcs*.

13. Clements, *Salmon at the Antipodes*, 41–51.

14. Boccius, *Fish in Rivers and Streams*, 26–27.

15. Neil Smith, "Sir James Arndell Youl (1811–1904)."

16. "Introduction of Salmon," *North Australian, Ipswich and General Advertiser*, 17 July 1860, 4. Reproduced from the *Launceston Examiner*, 19 June 1860, 4.

17. "Introduction of Salmon," 4.

18. Buckland, "On the Acclimatisation of Animals," 26.

19. *Command Papers, Reports Made for the Year 1859 to the Secretary of State*, 87–91.

20. Command Papers, *Reports Made for the Year 1859 to the Secretary of State*, 90.

21. Command Papers, *Reports Made for the Year 1859 to the Secretary of State*, 88.

22. Command Papers, *Reports Made for the Year 1859 to the Secretary of State*, 88. Black was later commissioned to investigate the possibility of acclimatizing Atlantic salmon in New South Wales. In August 1861, Black ventured into central New South Wales observing the current, inhabitants, and temperatures of local rivers. Based on the similarities between the Scottish, Tasmanian, and New South Wales temperatures, and the perceived absence of competing and predatory species in local rivers, Black recommended the introduction of nonmigratory salmon species, particularly the Danube salmon. "Parliamentary Paper," *SMH*, 25 November 1861, 2.

23. Minute Book One, 31–37.

24. Minute Book Two, 561–63. Brown trout and salmon trout are now seen to be two separate subspecies of *Salmo Trutta*.

25. Clements, *Salmon at the Antipodes*, 117–70.

26. "Fishing," *YA*, 5 September 1863, 774.

27. "The Salmon," *Ar*, 30 May 1864, 5.

28. Minard, "Assembling Acclimatization," 1–14.

29. Minute Book Two, 534–35; Report of the Tasmanian Salmon Commissioners, 1864, reproduced in Nichols, *Acclimatisation of the Salmonidae*, 134–64.

30. Minute Book Two, 563–66.

31. McCoy, "Zoology and Palaeontology," 187.

32. "The Acclimatisation Society," *Ar*, 3 June 1864, 5.

33. "Acclimatisation Society," 5.

34. Minute Book Two, 512–14.

35. Minute Book Two, 546–47.

36. Minute Book Two, 629–30.

37. Royal Society of Tasmania, *Papers and Proceedings*, 45–47.

38. The Tasmanian salmon commissioners and Albert Günther, from the British Museum, engaged in a long-running war of words over Tasmanian salmon. The commissioners believed that they has secured mature specimens of Atlantic salmon from the Derwent; Günther thought the specimens were just large brown trout, or, at best, trout-salmon hybrids. Clements, *Salmon at the Antipodes*, 95–98.

39. Minute Book Two, 688–90.

40. A.S.V., *Sixth Annual Report*, 1868.

41. Nichols, *Acclimatisation of the Salmonidae*, 201.

42. Clements, *Salmon at the Antipodes*, 100.

43. The 1864 Report of the Tasmanian Salmon Commissioners reproduced within Nichols, *Acclimatisation of the Salmonidae*, 201–2.

44. "Pisciculture," *Ar*, 19 September 1866, 5.

45. "Saturday, December 8, 1866," *Ar*, 8 December 1866, 7.

46. "Saturday, December 8, 1866," 7.

47. Z.A.S.V., *Proceedings and Report of the Annual Meeting*, 1874, 3: 38–44.

48. Clements, *Salmon at the Antipodes*, 170–71.

Chapter Six

1. Brett Stubbs has written an article on the history of the game laws in New South Wales, where he focused on the evolving attitudes toward native fauna and the contradiction between game acts that protected selected native animals and other acts that rewarded the destruction of native animals. He acknowledged the role of the ASNSW in establishing game laws in New South Wales, but thought its prime focus was on imported animals and that the protection of native animals was an afterthought. Additionally, he repeated the assertion that until the mid-1870s colonists thought that native wildlife offered a poor variety and quality of game. Norman and Young, "Short-Sighted and Doubly Short-Sighted"; Stubbs, "Useless Brutes," 27. The idea that Australian colonists saw the landscape as deficient in native animals suitable for game hunting has been exaggerated, especially with regard to wildfowl. Coral Dow has written an excellent article on sports hunting on the Gippsland lakes in Victoria in the 1870s. She argued that Victoria was considered to be incredibly rich in wildfowl and offered good prospects for game hunting. Dow, "Sportsman's Paradise," 145–64.
2. MacKenzie, *Empire of Nature*, 295–304.
3. See, for example, Brennan, "Imperial Game," 202–15.
4. Colpitts, *Game in the Garden*, 164–68; Hussain, "Sports-Hunting, Fairness and Colonial Identity," 112–26; Boyce, "Return to Eden," 287–307.
5. Hunter, "New Zealand Hunters in Africa," 483–501.
6. Griffin, *Blood Sport*, 97–110.
7. Thompson, *Whigs and Hunters*; Hopkins, *Long Affray*; Landry, *Invention of the Countryside*; Fisher, "Property Rights in Pheasants," 165–80.
8. Hopkins, *Long Affray*, 183–84.
9. Fisher, "Property Rights in Pheasants," 168.
10. Fisher, "Property Rights in Pheasants," 168; Hopkins, *Long Affray*, 197.
11. Hopkins, *Long Affray*, 199.
12. Hopkins, *Long Affray*, 225; Karsten, *Between Law and Custom*, 31–32.
13. *House of Commons Papers: Report from the Select Committee on the Game Laws 1845*, iv.
14. Fisher, "Property Rights in Pheasants," 171.
15. De Castella and Thornton-Smith, *Australian Squatters*, 93.
16. De Castella and Thornton-Smith, *Australian Squatters*, 92.
17. Westgarth, *Australia Felix*, 126–27.
18. Westgarth, *Colony of Victoria*, 452.
19. Westgarth, *Colony of Victoria*, 452.
20. Westgarth, *Colony of Victoria*, 452.
21. Westgarth, *Colony of Victoria*, 452.
22. Waterhouse, *Vision Splendid*, 25–26.
23. Waterhouse, *Vision Splendid*, 25–26.
24. Frost, "European Farming, Australian Pests," 129–43.
25. Ronald, *Hounds Are Running*, 5.
26. Ronald, *Hounds Are Running*, 5.
27. Sayers, *Letters from Victorian Pioneers*, 120.

28. Wheelwright, *Natural History Sketches*, 29.
29. Wheelwright, *Natural History Sketches*, 17.
30. Wheelwright, *Natural History Sketches*, 29.
31. Wheelwright, *Natural History Sketches*, 66.
32. Ostapenko, "Golden Horizons," 38–40.
33. Wheelwright, *Natural History Sketches*, 67.
34. "Protection to Game," *Ar*, 30 December 1864, 6.
35. Wheelwright, *Natural History Sketches*, 188.
36. Norman and Young, "Short-Sighted and Doubly Short-Sighted," 5.
37. *Act to Provide for the Preservation of Imported Game*, 1862, 3.
38. "An Australian Game Law," *Star*, 31 July 1860, 2.
39. "An Australian Game Law," *Star*, 31 July 1860, 2.
40. *Launceston Examiner*, 18 August 1860, as quoted in Bonyhady, *Colonial Earth*, 151.
41. *Launceston Examiner*, 18 August 1860, as quoted in Bonyhady, *Colonial Earth*, 151.
42. "Game Laws," *Cornwall Chronicle*, 31 August 1859, 4.
43. "Game Laws," *Cornwall Chronicle*, 31 August 1859, 4.
44. Stubbs, "Useless Brutes," 25. Similar acts were passed in South Australia in 1864, Queensland in 1863, and New South Wales in 1866.
45. Serle, *Golden Age*, 294–96.
46. Serle, *Golden Age*, 294–96.
47. Serle, *Golden Age*, 306; Lewis, "Snodgrass," 214–33.
48. Snodgrass overlanded from New South Wales in 1838 and established various pastoral holdings. He was a founding member of the Melbourne Cricket Club and the Melbourne Club. Politically he squarely aligned with the squatting interest. Lewis, "Snodgrass," 214–33.
49. Parliament of Victoria, *Victorian Hansard Third Parliament 1861*, 8: 231.
50. By this stage of his career, Don was in political strife because he spoke in favor of but eventually voted against Gavan Duffy's land act. His own supporters accused him of being bought off by the Victorian Association. Don may have been keen to reassert his radicalism by opposing the 1862 game act and to publicly clash with a prominent Victorian Association member like Snodgrass. Shiel, *Charles Jardine Don*, 158–62.
51. Legislative Council and Legislative Assembly, *Victorian Hansard Third Parliament 1861*, 8: 426.
52. Act to Provide for the Preservation of Imported Game, 1862, 3.
53. Act to Provide for the Preservation of Imported Game, 1862, 3.
54. Act to Provide for the Preservation of Imported Game, 1862, 1–4.
55. Minute Book One, 173–76, 155–60.
56. Minute Book One, 85–89.
57. News Clippings, 81.
58. Minute Book One, 91–93.
59. "Thursday, January 12," *Ar*, 12 January 1865, 4.
60. Fisher, "Property Rights in Pheasants," 169.
61. Act to Provide for the Preservation of Imported Game, 1862, 1–4.
62. Brennan, "Imperial Game," 102–6.
63. Rolls, *They All Ran Wild*, 25.

64. Griffin, *Blood Sport*, 49–55.
65. Wheelwright, *Natural History Sketches*, 186.
66. Wheelwright, *Natural History Sketches*, 187–88.
67. It is possible that brolgas were hunted for their feathers as part of the plumage trade, but I have found no evidence of this practice.
68. A.S.V., *Answers Furnished*, 8.
69. Rolls, *They All Ran Wild*, 271–330.
70. Letter Book Four, 672.
71. Rolls, *They All Ran Wild*, 214.
72. "The Melbourne Hunt," *Ar*, 16 July 1864, 6.
73. "Melbourne Hunt," 6.
74. "Temple Court in the 'Pink,'" *Ar*, 11 July 1864, 5.
75. "Temple Court in the 'Pink,'" 5.
76. "Temple Court in the 'Pink,'" 5.
77. Griffin, *Blood Sport*, 153–70.
78. Letter Book Three, 465.
79. "Melbourne," *Bendigo Advertiser*, 13 January 1863, 2.
80. "Victorian Game Laws," *Bendigo Advertiser*, 30 October 1863, 2.
81. *Act to Consolidate the Laws*, 1864. The ASV was routinely consulted about closed seasons and successfully advocated extended closed seasons in dry years. It was also successful in having native quail added to the list of protected native game. Minute Book Two, 579–82, 659–60, 855–56.
82. Waterhouse, *Vision Splendid*, 26. The Duffy Act was spectacularly unsuccessful, and loopholes were exploited by squatters to secure their land tenure. The amount of land under the control of squatters actually increased during the tenure of the Duffy Act.
83. "Acclimatisation Society's Dinner," *Ar*, 7 July 1864, 5.
84. "The Port Phillip Farmers Society," *Gippsland Times*, 15 July 1864, 3.
85. "Friday November 9, 1866," *McIvor Times and Rodney Advertiser*, 9 November 1866, 2.
86. "Friday November 9, 1866," 2.
87. "Protection to Game," *Ar*, 30 December 1864, 6.
88. Minute Book Two, 594–95.
89. "Parliament," *Ar*, 3 June 1865, 5.
90. "Alteration of the Close Season for Quail," *Ar*, 24 August 1866, 7.
91. Letter Book Five, 3, 12.
92. Minute Book Two, 705–6.
93. Minute Book Two, 705–6.
94. Z.A.S.V., *Proceedings and Annual Report 1872*, 1: 7.
95. Minute Book Two, 757–58.
96. *Act to Protect Game*, 1867, 126.
97. *Act to Protect Game*, 1867, 126.
98. Roberts, *History of Australian Land Settlement*, 253.
99. "Law Report," *Ar*, 5 November 1868, 6.
100. Archibald Michie was Victoria's first Queen's Counsel and one of the lawyers at the Eureka Stockade trial. He was periodically a member of Parliament throughout

the 1860s. He was attorney general in several colonial ministries and was a land reform advocate. He was not a member of Parliament while pursuing Dunn's case. Hall, "Michie, Sir Archibald (1813–1899)." George Higinbotham was a prominent liberal reformer, a land reform advocate, and attorney general in the McCulloch ministry at the time of the Dunn case. Dow, "Higinbotham, George (1826–1892)."

101. "Supreme Court Account of Dunn vs. Waldock," *Ar*, 5 November 1868, 6.
102. "The Farmers and the Hunt Club," *Ar*, 15 June 1869, 5.
103. "Farmers and the Hunt Club," 5.
104. Minute Book Two, 915–18.
105. Minute Book Three, 978–79.
106. "Deputations," *Ar*, 29 August 1871, 6.
107. "Deputations," 6.
108. "Deputations," 6.
109. Norman and Young, "Short-Sighted and Doubly Short-Sighted," 11.
110. "Swivel Gun Havoc," *Riverine Herald*, 25 June 1873, 2.
111. "The Swivel Gun Nuisance," *Ar*, 12 July 1873, 2.
112. "Swivel Gun Nuisance," 2.
113. "Swivel Gun Nuisance," 2.
114. "The Swivel Gun," *Ar*, 15 August 1873, 7.
115. Act to Further Amend an Act Intitled "An Act to Protect Game," 1873.
116. Act to Further Amend an Act Intitled "An Act to Protect Game," 1873.

Chapter Seven

1. Gillbank, "Animal Acclimatisation," 301–4; Dunlap, "Australian Nature," 28–30; Dunlap, "Remaking the Land," 311–16; Brennan, "Imperial Game," 135–37; Tyrrell, *True Gardens of the Gods*, 32–33. A historiographical tradition exists, originally established by one of the Le Souëfs' descendants, that argues that after the 1872 name change and the appointment of Albert Le Souëf, the ZASV deliberately turned away from acclimatization in order to focus on its new zoological collection. Le Souëf, "Development of a Zoological Garden," 221–52. This tradition was adopted by Dunlap, *Nature and the English Diaspora*, 57; Gillbank, "Animal Acclimatisation," 303–5; Brennan, "Imperial Game," 137–39.
2. A.S.V., *Sixth Annual Report*.
3. Minute Book Three, 980–82.
4. Minute Book Three, 927.
5. Minute Book Three, 1032.
6. De Courcy, *Evolution of a Zoo*.
7. McEvey, "Le Souef, Albert Alexander (1828–1902)."
8. De Courcy, *Evolution of a Zoo*.
9. De Courcy, *Evolution of a Zoo*.
10. In 1867, the ASV briefly considered starting a zoological collection when a group of exotic carnivores came on the market. It rejected the plan as impractical, expensive, and inconsistent with its aims. "Mr Stutt in Explanation," *Ar*, 18 May 1867, 6; Minute Book Three, 775–77, 781–83.

11. Minute Book Three, 948.
12. Minute Book Three, 948.
13. Minute Book Three, 1057–59.
14. Le Souëf, "Development of a Zoological Garden," 221–52.
15. "Acclimatisation Society of Victoria," *Ar*, 11 March 1871, 1; Z.A.S.V., *Proceedings and Report of the Annual Meeting*, 1872,:1; Z.A.S.V., *Proceedings and Report of the Annual Meeting*, 1873, 2.
16. Minute Book Three, 989–92.
17. Minute Book Three, 993–97.
18. Z.A.S.V., *Proceedings and Report of the Annual Meeting*, 1875, 4: 10.
19. "The Zoological and Acclimatisation Society," *Ar*, 1 March 1881, 6.
20. "Zoological Society," *Ar*, 5 February 1884, 9.
21. Moyal, *Bright & Savage Land*, 162–74; Finney, *Paradise Revealed*, 131–41; Powell, *Environmental Management in Australia*, 97–127; Hutton, *Australian Environment Movement*, 59–68; Bonyhady, *Colonial Earth*, 192–221.
22. Powell, *Watering the Garden State*, 104–20; Tyrrell, *True Gardens of the Gods*, 121–41; Frost, "Farmers, Government, and the Environment," 19.
23. For the definitive account of national nature in the English-speaking world, see Dunlap, *Nature and the English Diaspora*. For the history of bird protection in Britain, see Moore-Colyer, "Feathered Women and Persecuted Birds," 57–73; Bonhomme, "Nested Interests," 47. For an assessment of bird and mammal protection in the other Australian colonies, see Petrow, "Civilizing Mission," 71–94; Stubbs, "Useless Brutes," 25–50.
24. Alfred Newton was the University of Cambridge's first professor of zoology and comparative anatomy, and believed in two categories of extinction: natural caused by Darwinian evolution and unnatural caused by humanity. In 1868, Newton helped form the Closed Season Committee of the British Association for the Advancement of Science. By doing this, he hoped to establish closed seasons for indigenous British fauna, rationalize fauna protection based on extinction threat, and marginalize animal protectionists who wished to protect animals on anticruelty grounds. Cowles, "Victorian Extinction," 1–22. Alfred Russel Wallace is primarily remembered as the codiscoverer of evolution by natural selection and as founder of the modern science of biogeography. He argued for the centrality of competition and physical isolation in determining the evolution and distribution of organisms. Wallace was also a socialist, a spiritualist, and sometimes a critic of empire. Endersby, "Escaping Darwin's Shadow," 385–401; Clark and York, "Restoration of Nature and Biogeography," 216–30.
25. Houghton and Presland, *Leaves from Our History*, 5.
26. Houghton and Presland, *Leaves from Our History*, 5.
27. Houghton and Presland, *Leaves from Our History*, 5.
28. Mulvaney, "Spencer, Baldwin (1860–1929)."
29. Mulvaney and Calaby, "*So Much That Is New*," 79, 149–50.
30. Barrow, *Nature's Ghosts*, 39–46.
31. Cowles, "Victorian Extinction," 4–7.
32. Warwick Anderson has researched Wallace's understanding of acclimatization, but did not draw out its implications for Australia, nor link it to a broader scientific

critique of introduced species in the late nineteenth century. Anderson, "Climates of Opinion," 151. Matthew Chew has connected Wallace to this critique. He did not, however, link Wallace's acclimatization papers to Wallace's broader oeuvre. Chew, "Ending with Elton," 64.

33. *Encyclopaedia Britannica*, 9th ed. s.v. "Acclimatisation."
34. *Encyclopaedia Britannica*, 9th ed. s.v. "Acclimatisation."
35. *Encyclopaedia Britannica*, 9th ed. s.v. "Acclimatisation."
36. Chew, "Ending with Elton," 67.
37. Wallace, *Darwinism*, 34.
38. Competitive exclusion is still discussed in ecological literature today. See, for example, McPeek, "Limiting Factors," 3–5.
39. Wallace, *Darwinism*, 110.
40. Diez et al., "Darwin's Naturalization Conundrum," 674–81; Thuiller et al., "Darwin's Naturalization Conundrum," 461–75.
41. Star, "Plants, Birds and Displacement Theory," 12–16; Star, "From Acclimatisation to Preservation," 271–79.
42. "Australian Animals: Their Past and Present History," *Ar*, 25 August 1888, 5.
43. "Australian Animals: Their Past and Present History," *Ar*, 25 August 1888, 5.
44. "Australian Animals: Their Past and Present History," *Ar*, 8 September 1888, 6.
45. "Australian Animals: Their Past and Present History," *Ar*, 8 September 1888, 6.
46. "Australian Animals: Their Past and Present History," *Ar*, 8 September 1888, 6.
47. "Australian Animals: Their Past and Present History," *Ar*, 25 August 1888, 5.
48. Mulvaney and Calaby, "So Much That Is New," 259–61.
49. This narrative is articulated in Norman and Young, "Short-Sighted and Doubly Short-Sighted," 2–13; Hutton and Connors, *Australian Environment Movement*, 40–43; Bonyhady, *Colonial Earth*, 130–35, 212–14.
50. Archibald Campbell (1853–1921) was a dedicated ornithologist and conservationist. Campbell helped create the (Royal) Australasian Ornithologists' Union; he was president in 1909 and 1928 and coeditor of its journal, the *Emu*, for thirteen years. In addition, he was an honorary associate in ornithology at the National Museum, and lectured in nature study to the Workers' Educational Association. He published within *Southern Science Record*, the *Victorian Naturalist*, and the *Proceedings of the Royal Society of Victoria*.
51. Campbell, "Protection of Native Birds," 161.
52. Campbell, "Protection of Native Birds," 161.
53. Minute Book Five, 6, 56.
54. Minute Book Five, 60.
55. Minute Book Five, 60.
56. "Monthly Meeting, May 1885."
57. *News Clippings re Zoological and Acclimatisation Society*, 43.
58. Campbell, "Protection of Native Birds," 161.
59. Campbell, "Protection of Native Birds," 162.
60. Matthams, *Rabbit Pest in Australia*; Rolls, *They All Ran Wild*; Coman, *Tooth & Nail*, 431–57; Dando-Collins, *Pasteur's Gambit*; Noble and Pfitzner, "They Know Not What They Do," 431–57.

61. Coman, *Tooth & Nail*, 20–25.

62. Dunlap, *Nature and the English Diaspora*, 80–90.

63. "The Distribution of Animals," *Ti*, 29 August 1860, 11; "Acclimatisation," *SMH*, 15 December 1862, 3; Minard, "Assembling Acclimatization," 8.

64. Minute Book Two, 592–93, 619–620, 647–49.

65. Minute Book Two, 408–11.

66. The *Land and Water* article was reproduced as "Acclimatisation in Victoria," *Queenslander*, 5 September 1868, 3.

67. Minute Book Three, 912–13.

68. "Australian Rabbits," *Launceston Examiner*, 23 July1870, 3. This article quotes Frank Buckland's Salmon Commission reports as discussing displacement of English rats by Norwegian black rats, discussing rabbits in Australia, and placing these displacements in the context of the struggle for existence. This is somewhat perplexing, as Buckland was the president of SAUK and ardently anti-Darwinian. Collins, "From Anatomy to Zoophagy," 92–93.

69. "The Royal-Park Gardens," *Ar*, 5 December 1876, 6.

70. Parliament of Victoria, *Votes and Proceedings of the Legislative Assembly 1884*.

71. "Parliament," *Ar*, 18 September 1884, 9.

72. "Sunday June 21 1884," *Ar*, June 21 1884, 8.

73. Dunlap, *Nature and the English Diaspora*, 87.

74. Star, "From Acclimatisation to Preservation," 259–60.

75. Wells, "Enemy of the Rabbit," 301–3.

76. Wells, "Enemy of the Rabbit," 301–3.

77. "Destruction of Rabbits," *Otago Witness*, 28 October 1876, 18.

78. Cowles, "Victorian Extinction," 3–4.

79. "The Mongoose as a Rabbit Destroyer," *Ar*, 5 October 1881, 10.

80. "Mongoose as a Rabbit Destroyer," 10.

81. "The Mongoose as a Rabbit Destroyer," *Ar*, 27 May 1884, 7; "The Mongoose," *Ar*, 16 August 1884, 5.

82. "The Mongoose for Australia," *Camperdown Chronicle*, 28 February 1883, 4.

83. "The Mongoose," *Ar* 16 August 1884, 5.

84. "Mongoose," 5.

85. "Parliament," *Ar*, 18 September 1884, 9.

86. Rolls, *They All Ran Wild*, 113–50; Dando-Collins, *Pasteur's Gambit*, 64–79.

87. In France and New Zealand, prominent acclimatizers-turned-conservationists also wrote articles skeptical of acclimatization. See Osborne, *Nature, the Exotic*, 52–57; Star, "Plants, Birds and Displacement Theory," 14–16.

88. Le Souëf, "Acclimatisation in Victoria," 476.

Chapter Eight

1. Griffiths, *Hunters and Collectors*, 123–25.

2. Taylor, *Making Salmon*, 68–99; Towle, "Authored Ecosystems," 58–65; Halverson, *Entirely Synthetic Fish*, 28–48; MacLeod, "Government and Resource Conservation," 114–50; Kinsey, "Seeding the Water," 527–66.

3. Clements, *Salmon at the Antipodes*, 170–71.

4. Taylor, *Making Salmon*, 72–76.

5. Hutton, *Australian Environment Movement*, 63–65.

6. Griffiths, *Hunters and Collectors*, 123–36.

7. For the history of fly-fishing in Australia, see Dunn, *Angling*, 61–97; Clark, *Catch*, 65–75. For a discussion of the distinctiveness of Australian fly-fishing, please see Franklin, "Australian Hunting and Angling Sports," 39–56; Franklin, "Performing Acclimatisation," 19–44.

8. Franklin, "Australian Hunting and Angling Sports," 39–56; Franklin, "Performing Acclimatisation," 19–44.

9. Clements, *Salmon at the Antipodes*, 217. Robert de Bruce Johnstone served on the Geelong Council for twenty-three years (he was mayor three times) and was member of the Legislative Assembly for twelve years. During this whole period, he maintained a successful saddlery business in Geelong. John Raddenberry was the second curator of the Geelong Botanical Gardens. He served for over thirty years and extensively redesigned the gardens for public recreation.

10. Connor Papers, *Fish Acclimatisation*.

11. Connor Papers, *Rules of the Geelong and Western District Fish Acclimatising Society*.

12. Connor Papers, *Rules of the Geelong and Western District Fish Acclimatising Society*.

13. Connor Papers, *Fish Acclimatisation*, 1.

14. Clements, *Salmon at the Antipodes*, 181.

15. Sir Samuel Wilson was a founding member of the ASV, and an active member of the regional fish acclimatization societies. He experimented with ostrich farming, sericulture, and raising angora goats.

16. Wilson, *Californian Salmon*, 22.

17. Wilson, *Californian Salmon*, 26.

18. Lever, *Naturalized Fishes*, 35. No more attempts were made to introduce Californian salmon to Victoria until the state Department of Fisheries and Game began stocking lakes with salmon in the 1930s. Today a small population of Californian salmon is maintained by artificial stocking in landlocked lakes in rural Victoria. No self-sustaining populations of Californian salmon have ever been established in Victoria.

19. "The Angler's Society," *Ar*, 4 November 1875, 10.

20. Z.A.S.V., *Proceedings and Report of the Annual Meeting*, 1872, 1: 171.

21. Senior, *Travel and Trout*, 76.

22. Samuel Wilson, *Salmon at the Antipodes*, 184.

23. Clements, *Salmon at the Antipodes*, 180–90, 220–22.

24. Robinson, "Evolution of Some Key Elements," 129–50; Taylor, *Making Salmon*; Towle, "Authored Ecosystems," 54–74.

25. For professionalized forestry policy in Australia and the world, see Powell, *Environmental Management in Australia*; Barton, "Empire Forestry," 529–52. For studies of fisheries management, see Jacobsen, "Steam Trawling"; Halverson, *Entirely Synthetic Fish*.

26. "Tasmanian Fisheries," *Mercury*, 21 October 1884, 3; Saville-Kent, *Observations*.

27. Harrison, *Savant of the Australian Seas*; Harrison, "Fisheries Savant"; McCalman, *Reef*, 168–96.

28. Parliament of Victoria, *Votes and Proceedings, General Report on the Fisheries*, 1888.

29. Parliament of Victoria, *Votes and Proceedings, General Report on the Fisheries*, 1888.

30. Parliament of Victoria, *Votes and Proceedings, General Report on the Fisheries*, 1888, 26.

31. Parliament of Victoria, *Votes and Proceedings, General Report on the Fisheries*, 1888.

32. Parliament of Victoria, *Votes and Proceedings, General Report on the Fisheries*, 1888, 26.

33. "Fish Propagation in Victoria," *Aust*, 7 July 1888, 52.

34. Clements, *Salmon at the Antipodes*, 231–35.

35. Powell, *Watering the Garden State*, 91–27; Waterhouse, *Vision Splendid*, 86–90.

36. The Piscatorial Council of Victoria (formed in 1904) was an alliance of twenty-five angling and fish acclimatization societies that advocated an interesting combination of developing fisheries science and the promotion of aquaculture and recreational fishing. "Angling," *Ars*, 22 October 1904, 18.

37. Lever, *Naturalized Fishes*, 27–28.

38. Clements, *Salmon at the Antipodes*, 140–42.

39. Macdonald, *Gum Boughs*, 101.

40. "Trout Hatching," *Ar*, 9 March 1910, 4; "Acclimatisation Society," *Ar*, 5 March 1910, 21.

41. "Victorian Fisheries,' *Ar*, 23 April 1907, 9; "Nature Notes and Queries," *Ar*, 15 July 1910, 5; "Nature Notes and Queries" *Ar*, 7 January 1910, 9.

42. Saville-Kent, *Naturalist in Australia*, 156; Macdonald, *Gum Boughs*, 93.

43. "Nature Notes and Queries," *Ar*, 11 February 1916, 8.

44. "Nature Notes and Queries," *Ar*, 25 February 1916, 5.

45. Fisheries and Wildlife Division, General Correspondence Subject Files.

46. Fisheries and Wildlife Division, General Correspondence Subject Files.

47. "Our Native Fish," *Traralgon Record*, 2 May 1916, 4.

48. "The Twelve Best Birds," *Ar*, 24 October 1908, 19.

Epilogue

1. Mueller, "Transactions of the Philosophical Institute," 3.

Bibliography

Primary Sources

Manuscripts

Geelong, Victoria
  Geelong Heritage Centre
    JH Connor Papers 373/1/1-2
    Geelong and Western District Fish Acclimatising Society
Melbourne, Victoria
  Public Record Office Victoria
    Acclimatisation Society of Victoria Letter Books Outwards, VPRS 2225/
      P0000/1-5
    Letter Books One to Five
Melbourne, Victoria
  Public Record Office Victoria
    Acclimatisation Society of Victoria Minute Books, VPRS 2223/P0000/1-6
    Minute Books One to Six
Melbourne, Victoria
  Public Record Office Victoria
    Fisheries and Wildlife Division, VPRS 12011/0001/1
      General Correspondence Subject Files
Melbourne, Victoria
  Public Record Office Victoria
    News Clippings re Zoological and Acclimatisation Society 01 Jan 1869 to 1910
    Records of the Royal Melbourne Zoological Gardens and Related Organisations,
      VPRS 8850/P00001/81
Melbourne, Victoria
  State Library of Victoria
    Australian Joint Copying Project Miscellaneous Series
    Owen Papers, 2278/18

Printed Sources

Acclimatisation Society of New South Wales. *The Third Annual Report of the Acclimatisation Society of New South Wales: With an Address on the Physiology, Utility, and Importance, of Acclimatisation, to Australia; Delivered at the Annual Meeting of the Society, April 4th, 1864.* Sydney: Printed by Joseph Cook, 1864.

Acclimatisation Society of Victoria. *Answers Furnished by the Acclimatisation Society of Victoria to the Enquiries Addressed to It by His Excellency the Governor of Victoria at the*

Instance of the Right Hon. The Secretary of States for the Colonies. Melbourne: Wilson and MacKinnon, 1864.

———. *The Fifth Annual Report of the Acclimatisation Society of Victoria as Adopted at the Annual Meeting of the Society, Held February 22nd, 1867, at the Mechanics' Institute, Melbourne.* Melbourne: Wilson and Mackinnon, 1867.

———. *The First Annual Report of the Acclimatisation Society of Victoria with the Addresses Delivered at the Annual Meeting of the Society Held November 24th, 1862, at the Mechanics Institute, Melbourne, by His Excellency Sir Henry Barkly, K.C.B. and Professor McCoy.* Melbourne: F. A. Masterman, 1862.

———. *The Fourth Annual Report of the Acclimatisation Society of Victoria.* Melbourne: Wilson and Mackinnon, 1865.

———. *The Rules and Objects of the Acclimatisation Society of Victoria, with the Report Adopted at the First General Meeting of the Members, and a List of the Officers, Members and Subscribers to the Society.* Melbourne: William Goodhugh, 1861.

———. *The Second Annual Report of the Acclimatisation Society of Victoria with the Address Delivered at the Annual Meeting of the Society, Held November 11th, 1863 at the Society's Office, Melbourne, by His Excellency Sir C. H. Darling, K.C.B.* Melbourne: F. A. Masterman, 1863.

———. *The Sixth Annual Report of the Acclimatisation Society of Victoria.* Melbourne: Wilson and Mackinnon, 1868.

———. *The Third Annual Report of the Acclimatisation Society of Victoria as Adopted at the Annual Meeting of the Society, Held November 11, 1864 at the Society's Office, Melbourne — Together with Papers Read at the Monthly Meetings of the Society.* Melbourne: Wilson and Mackinnon, 1864.

Ashworth, Thomas. *Essay on the Practical Cultivation of a Salmon Fishery, Addressed to the President and Council of the International Congress, to Promote the Cultivation of Fisheries, Held at Arcachon, 1866.* London: Judd and Glass, 1866.

Ashworth, Thomas, and Edward Ashworth. *A Treatise on the Propagation of Salmon and Other Fish.* London: E. H. King, 1854.

Bennett, G. "Acclimatisation: Its Eminent Adaptation to Australia; A Lecture Delivered in Sydney, by George Bennett F.R.S., &C. &C." Melbourne: Acclimatisation Society of Victoria, 1862.

———. *Gatherings of a Naturalist in Australasia: Being Observations Principally on the Animal and Vegetable Productions of New South Wales, New Zealand and Some of the Austral Islands.* Milson's Point: Currawong, 1982.

———. *On the Introduction, Cultivation, and Economic Uses of the Orange and Others of the Citron-Tribe in New South Wales.* Sydney: New South Wales Government, 1871.

———. *Wanderings in New South Wales, Batavia, Pedir Coast, Singapore and China: Being the Journal of a Naturalist in Those Countries during 1832, 1833 and 1834.* Facsimile Edition. 2 vols. Vol. 1. Adelaide: Libraries Board of South Australia, 1967.

Blyth, Edward. *Catalogue of the Mammalia in the Museum Asiatic Society.* Calcutta: Savielle and Cranenburgh, 1863.

———. *Report on the Mammalia and More Remarkable Species of Birds Inhabiting Ceylon.* Ceylon: The Author, 1851.

Boccius, Gottlieb. *Fish in Rivers and Streams: A Treatise on the Production and Management of Fish in Fresh Waters, by Artificial Spawning, Breeding and Rearing: Showing Also the Cause of the Depletion of All Rivers and Streams.* London: John van Voorst, 1848.

Buckland, Frank. *Natural History of British Fishes: Their Structure, Economic Uses and Capture by Net and Rod; Cultivation of Fish Ponds, Fish Suited for Acclimatisation.* London: Society for Promoting Christian Knowledge, 1880.

———. "On the Acclimatisation of Animals." *Journal of the Society of the Arts* 9 (1860): 19–35.

Butler, E. A., and A. O. Hume. *A Catalogue of the Birds of Sind, Cutch, Káthiáwár, North Gujarát, and Mount Aboo . . . With References (Confined as Much as Possible to Jerdon's Birds of India, Mr. Hume's Raptores and Stray Feathers) Showing Where Each Species Is Described . . . Contributed to the Bombay Gazetteer.* Bombay: Bombay Gazetteer, 1879.

Campbell, A. J. "Protection of Native Birds." In *The Victorian Naturalist*, edited by the Field Naturalists' Club of Victoria, 164–67. Melbourne: Field Naturalists' Club of Victoria, 1884.

Cuvier, Georges, and Edward Griffith. *The Class Pisces Arranged by Baron Cuvier with Supplementary Additions by Edward Griffith F. R. S., &C. and Lieut-Col Charles Hamilton Smith, C. H., K. W., F. R., L. S. S., &C. &C.* London: Whittaker, 1834.

Cuvier, Georges, and Achille Valenciennes. *Histoire Naturelle Des Poissons.* 22 vols. Vol. 18. Paris: P. Bertrand, 1846.

Davey, H. W. "Upsetting the Balance of Nature." *Victorian Naturalist* 33, no. 5 (1917): 151–54.

De Castella, Hubert, and C. B. Thornton-Smith. *Australian Squatters.* Carlton: Melbourne University Press, 1987.

"Destruction of Small Birds by Farmers, Gardeners and Others." *Zoologist: A Popular Monthly Magazine of Natural History* 214 (1861): 7729–40.

"Fisheries of Victoria." *Public Lands Circular* 10 (1862).

Günther, Albert. *Catalogue of the Physostomi, Containing the Families Salmonidae, Percopsidae, Galaxidae, Mormyridae, Gymnarchidae, Esocidae, Umbridae, Scombresocidae, Cyprinodontidae in the Collection of the British Museum.* 10 vols. Vol. 6 of *Catalogue of the fishes in the British Museum.* London: British Museum Trustees, 1866.

Günther, Albert. *Ichthyology, Encyclopaedia Britannica.* 9th edition. 24 vols. Vol. 12. Edinburgh: Adam and Charles Black, 1881.

Jerdon, T. C. *The Birds of India, Being a Natural History of All the Birds Known to Inhabit Continental India: With Descriptions of the Species, Genera, Families, Tribes, and Orders, and a Brief Notice of Such Families as Are Not Found in India, Making It a Manual of Ornithology Specially Adapted for India.* Calcutta: Military Orphan, 1863.

Layard, E. L. *Catalogue of the Specimens in the Collection of the South African Museum.* Part 1, *The Mammalia.* Cape Town: Cape Town Argus, 1862.

———. "XVIII.—Notes on the Ornithology of Ceylon, Collected during an Eight Years' Residence in the Island." *Annals and Magazine of Natural History* 12, no. 69 (1853): 165–76.

———. "V.—Notes on the Ornithology of Ceylon, Collected during an Eight Years' Residence in the Island." *Annals and Magazine of Natural History* 14, no. 79 (1854): 57–64.

———. "XLI.—Notes on the Ornithology of Ceylon, Collected during an Eight Years' Residence in the Island." *Annals and Magazine of Natural History* 13, no. 78 (1854): 446–53.

———. "X.—Notes on the Ornithology of Ceylon, Collected during an Eight Years' Residence in the Island." *Annals and Magazine of Natural History* 12, no. 68 (1853): 97–107.

———. "XIII.—Notes on the Ornithology of Ceylon, Collected during an Eight Years' Residence in the Island." *Annals and Magazine of Natural History* 13, no. 74 (1854): 123–31.

———. "XXV.—Notes on the Ornithology of Ceylon, Collected during an Eight Years' Residence in the Island." *Annals and Magazine of Natural History* 14, no. 82 (1854): 264–72.

———. "XXIV.—Notes on the Ornithology of Ceylon, Collected during an Eight Years' Residence in the Island." *Annals and Magazine of Natural History* 13, no. 76 (1854): 257–64.

———. "XXI.—Notes on the Ornithology of Ceylon, Collected during an Eight Years' Residence in the Island." *Annals and Magazine of Natural History* 13, no. 75 (1854): 212–18.

Le Souëf, W. H. D. "Acclimatisation in Victoria." In *Report of the Second Meeting of the Australasian Association for the Advancement of Science*, edited by Baldwin Spencer, 476–82. Melbourne: Australasian Association for the Advancement of Science, 1890.

Macdonald, Donald. *Gum Boughs and Wattle Bloom, Gathered on Australian Hills and Plains.* London: Cassell, 1887.

Marsh, George Perkins. *Man and Nature or, Physical Geography as Modified by Human Action.* London: Charles Scribner, 1864.

Matthams, James. *The Rabbit Pest in Australia: With Chapters on Foxes, Dingoes, Wombats, the Fences Act of Victoria and Noxious Weeds.* Melbourne: Speciality Press, 1921.

McCoy, Frederick. "Acclimatisation of Animals." *Edinburgh Review* 111, no. 25 (1860): 160–80.

———. "Note on the Ancient and Recent Natural History of Victoria." *Annals and Magazine of Natural History* 9, no. 3 (1861): 137–51.

———. "On the Recent Zoology and Palaeontology of Victoria." *Annals and Magazine of Natural History* 20, no. 107 (1865): 175–203.

———. *The Order and Plan of Creation: The Substance of Two Lectures Delivered in Connection with the Early Closing Association.* Melbourne: Samuel Mullen, 1870.

Mitchell, David William. "Acclimatisation of Animals." *Edinburgh Review* 111, no. 25 (1860): 160–80.

"Monthly Meeting, May 1885." In *The Victorian Naturalist*, edited by Field Naturalists' Club of Victoria, 1. Melbourne: Field Naturalists' Club of Victoria, 1885.

Mueller, Ferdinand. "Address of the President, Ferdinand Mueller, Delivered to the Members of the Philosophical Institute at the Inauguration Meeting in the New Hall." Melbourne: Philosophical Institute of Victoria, 1860.

———. "The Application of Phytology to the Industrial Purposes of Life." Melbourne: Samuel Mullen, 1871.

———. "Forest Culture in Its Relation to Industrial Pursuits: A Lecture." Melbourne: Victorian Government Printer, 1871.

———. "Transactions of the Philosophical Institute of Victoria: Anniversary Address of the President, Ferdinand Mueller, Esq., Ph.D., M.D., F.R.G.S., &C., &C., Government Botanist for the Colony of Victoria." Melbourne: Philosophical Institute of Victoria, 1859.

Nichols, Arthur. *The Acclimatisation of the Salmonidae at the Antipodes: Its History and Results.* London: Sampson Low, Marson Searle and Rivington, 1882.

Palmer, T. S. *The Danger of Introducing Noxious Animals and Birds: Reprint from Yearbook of Department of Agriculture 1893.* Washington, DC: Department of Agriculture, 1895.

Royal Society of Tasmania. *Papers and Proceedings of the Royal Society of Tasmania.* Hobart: Royal Society of Tasmania, 1866.

Saville-Kent, William. *The Naturalist in Australia.* London: Chapman and Hall, 1897.

———. *Observations on the Acclimatisation of the True Salmon (Salmo Salar), in Tasmania Waters, and upon the Reported Salmon Disease at the Breeding Establishment on the River Plenty.* Hobart: Royal Society of Tasmania, 1885.

Sayers, C. E, ed. *Letters from Victorian Pioneers, Being a Series of Papers on the Early Occupation of the Colony, the Aborigines Etc. Addressed by Victorian Pioneers to His Excellency Charles Joseph Latrobe, Esq. Lieutenant-Governor of the Colony of Victoria.* Melbourne: Thomas Francis Bride, 1969.

Senior, William. *Travel and Trout in the Antipodes—an Angler's Sketches in Tasmania and New Zealand.* London: Chatto and Windus, 1880.

Smith, James. *The Cyclopedia of Victoria: An Historical and Commercial Review: Descriptive and Biographical, Facts, Figures and Illustrations: An Epitome of Progress.* Melbourne: Cyclopedia, 1902.

Thomson, George M. *The Naturalisation of Animals & Plants in New Zealand.* Cambridge: Cambridge University Press, 1922.

Wallace, Alfred Russel. *Darwinism: An Exposition of the Theory of Natural Selection with Some of Its Applications.* London: Macmillan, 1890.

Walton, William. *The Alpaca: Its Naturalisation in the British Isles Considered a National Benefit, and an Object of Immediate Utility to the Farmer and Manufacturer.* New York: Office of the New York Farmer and Mechanic, 1845.

Westgarth, William. *Australia Felix, or a Historical and Descriptive Account of the Settlement of Port Phillip, New South Wales: Including Full Particulars of the Manners and Customs of the Aboriginal Natives.* Edinburgh: Oliver and Boyd, 1848.

———. *The Colony of Victoria: Its History, Commerce and Gold Mining; Its Social and Political Institutions Down to the End of 1863; with Remarks, Incidental and Comparative, upon the Other Australian Colonies.* London: Sampson Low, Son, and Marston, 1864.

Wheelwright, H. W. *Natural History Sketches. By the Old Bushman. Author of "Sporting Sketches, Home and Abroad." Etc.* London: Frederick Warne, 1864.

Wilson, Edward. *Acclimatisation: Read before the Royal Colonial Institute.* London: Royal Colonial Institute, 1875.

———. *Rambles at the Antipodes: A Series of Sketches of Moreton Bay, New Zealand, the Murray River and South Australia, and the Overland Route.* London: W. H. Smith and Son, 1859.

Wilson, Samuel. *The Californian Salmon, with an Account of Its Introduction into Victoria.* Melbourne: Sands and McDougall, 1878.

———. *Report on the Ostriches, Belonging to the Zoological and Acclimatisation Society, Now Running at Longerenong: Also Some Letters and Other Information about the Management of This Bird in South Africa.* Melbourne: Acclimatisation Society of Victoria, 1873.

———. *Salmon at the Antipodes: Being an Account of the Successful Introduction of Salmon and Trout into Australian Waters.* London: Edward Stanford, 1879.

Zoological and Acclimatisation Society of Victoria. *Proceedings of the Zoological and Acclimatisation Society of Victoria and Report of the Annual Meeting of the Society, Held 1st March, 1872.* 2 vols. Melbourne: Wilson and Mackinnon, 1872.

———. *Proceedings of the Zoological and Acclimatisation Society of Victoria and Report of the Annual Meeting of the Society, Held on the 24th February, 1873.* 2 vols. Melbourne: Wilson and Mackinnon, 1873.

———. *Proceedings of the Zoological and Acclimatisation Society of Victoria and Report of the Annual Meeting of the Society, Held 23rd February, 1874.* 3 vols. Melbourne: Wilson and Mackinnon, 1874.

———. *Proceedings of the Zoological and Acclimatisation Society of Victoria and the Report of the Annual Meeting of the Society, Held 24th February, 1875.* 4 vols. Melbourne: Wilson and Mackinnon, 1875.

Zoological Society of London. *Annual Report of the Zoological Society of London: Read at the Annual General Meeting, April 29 1861.* London: Zoological Society of London, 1861.

Newspapers

New South Wales
  Sydney Morning Herald, 1862–64
New Zealand
  Otago Witness, 1876
Queensland
  North Australian, Ipswich and General Advertiser, 1860
  Queenslander, 1868
Tasmania
  Launceston Examiner, 1858–70
  Mercury, 1884
United Kingdom
  Field, the Country Gentleman's Newspaper, 1860–62

   Gentleman's Magazine, 1868
   Glasgow Herald, 1860
   The Times, 1858–60
Victoria
   Argus, 1856–1916
   Australasian, 1864–88
   Ballarat Star, 1860–63
   Bendigo Advertiser, 1863–91
   Camperdown Chronicle, 1853
   Courier, 1862
   Gippsland Times, 1864
   McIvor Times and Rodney Advertiser, 1866
   Melbourne Punch, 1863
   North Melbourne Courier and West Melbourne Advertiser, 1908
   Portland Guardian and Normanby General Advertiser, 1856
   Riverine Herald, 1873–74
   Traralgon Record, 1916
   Yeoman and Australian Acclimatiser, 1863–64

Government Documents

Parliament of Great Britain. *Command Papers, Report of the Commissioners Appointed to Inquire into the Salmon Fisheries (England and Wales);Together with Minutes of Evidence*. No. 2768-1. London: House of Commons, 1861.

———. *Command Papers, Reports Made for the Year 1859 to the Secretary of State Having the Department of the Colonies; in Continuation of the Reports Annually Made by the Governors of the British Colonies, with a View to Exhibit Generally the Past and Present State of Her Majesty's Colonial Possessions. Transmitted with the Blue Books for the Year 1859*. Vol. 2. London: House of Commons, 1859.

———. *House of Commons Papers: Reports of Committees, 1845, Report from the Select Committee on the Game Laws*, 463-I. London: House of Commons, 1846.

Parliament of Victoria. *Parliamentary Debates / Legislative Council and Legislative Assembly, 1862 Session One, The Victorian Hansard Third Parliament*. Vol. 8. Melbourne: Victorian Government Printing Office, 1861.

———. *Votes and Proceedings of the Legislative Assembly 1884, Select Committee of the Legislative Assembly upon the Zoological and Acclimatisation Society of Victoria Incorporation*. Melbourne: Victorian Government Printing Office, 1884.

———. *Votes and Proceedings of the Legislative Assembly 1858–1859, Fisheries of the Colony*, E No.7. Vol. 1, Part 2. Melbourne: Victorian Government Printing Office, 1858.

———. *Votes and Proceedings of the Legislative Assembly 1862–1863, Session One, Petition from Certain Fishermen*. Vol. 1. Melbourne: Victorian Government Printing Office, 1863.

———. *Votes and Proceedings of the Legislative Assembly with Copies of Various Documents Ordered by the Assembly to Be Printed, General Report on the Fisheries of*

Victoria, by W. Saville Kent, F.L.S, F.Z.S., C. No 2. Vol. 1. Melbourne: Victorian Government Printing Office, 1888.

———. *Votes and Proceedings of the Legislative Assembly Session 1892-1893, The Fisheries Inquiry Board Report*, 878. Vol. 2. Melbourne: Victorian Government Printing Office, 1892.

———. *Votes and Proceedings of the Legislative Assembly 1908-1909, Second Session, the Fisheries Inquiry Board Report*. Vol. 1. Melbourne: Victorian Government Printing Office, 1908.

Statutes

An Act for the Preservation of Fish in the Lakes and Rivers of the Colony of Victoria, 1859, 22 Victoria.
An Act for the Protection of the Fisheries of Victoria, 1859, 22 Victoria.
An Act for the Regulation of the Oyster Fisheries in Victoria, 1859, 22 Victoria.
An Act to Amend and Consolidate the Laws for the Protection of the Fisheries of Victoria, 1862, 25 Victoria.
An Act to Amend the Laws Relating to Fisheries of Salmon in England, 1861, 24 and 25.
An Act to Carry into Effect a Convention between Her Majesty and the King of the French Concerning the Fisheries in the Seas between the British Islands and France, 1843, 6 and 7 Victoria.
An Act to Consolidate the Laws for the Protection of Fisheries and Game, 1864, 27 Victoria.
An Act to Further Amend an Act Intituled "an Act to Protect Game," and to Repeal Act No. 328, 1873, 37 Victoria.
An Act to Protect Game, 1867, 30 Victoria.
An Act to Provide for the Preservation of Imported Game and during the Breeding Season of Native Game, 1862, 25 Victoria.

Secondary Sources

Anderson, Warwick. "Climates of Opinion: Acclimatisation in Nineteenth-Century France and England." *Victorian Studies* 35, no. 2 (1992): 135–67.
Andrews, A. B. "Ledger, Charles (1818-1905)." *Australian Dictionary of Biography Online*. Accessed 25 August 2018. http://adbonline.anu.edu.au/biogs/A050085b.htm?hilite=ledger.
Australian Medical Pioneers Index. "Henry Ridgewood Madden." Barwon Health–Geelong Hospital Library. Accessed 25 August 2018. http://www.medicalpioneers.com/cgi-bin/index.cgi?detail=1&id=1224.
Ballantyne, Tony. *Orientalism and Race: Aryanism in the British Empire*. Basingstoke: Palgrave Macmillan, 2001.
Barrow, Mark B. *Nature's Ghosts: Confronting Extinction from the Age of Jefferson to the Age of Ecology*. Chicago: University of Chicago Press, 2009.

Barton, Gregory. "Empire Forestry and the Origins of Environmentalism." *Journal of Historical Geography* 27, no. 4 (2001): 529–52.
Beattie, James. "Acclimatisation and the 'Europeanisation' of New Zealand, 1830s–1920s?" *Environment and Nature in New Zealand* 3, no. 1 (2008): 100–20.
———. *Empire and Environmental Anxiety—Health Science, Art and Conservation in South Asia and Australasia, 1800-1920*. London: Palgrave Macmillan, 2011.
———. "The Empire of the Rhododendron: Reorienting New Zealand Garden History." In *Making a New Land: Environmental Histories of New Zealand*, edited by Tom Brooking and Eric Pawson, 241–61. Dunedin: Otago University Press, 2013.
———. "Plants, Animals and Environmental Transformation: New Zealand/Indian Biological and Landscape Connections, 1830s –1890s." In *The East India Company and the Natural World*, edited by Vinita Damodaran and Anna Winterbotham, 219–48. Basingstoke: Palgrave Macmillan, 2014.
Beinart, William. *The Rise of Conservation in South Africa: Settlers, Livestock and the Environment 1770-1950*. Oxford: Oxford University Press, 2003.
Bennett, Bruce. *Fish Markets of Melbourne*. Hawthorn: B. Bennett, 2002.
Bonhomme, Brian. "Nested Interests: Assessing Britain's Wild-Bird-Protection Law of 1869–1880." *Nineteenth-Century Studies* 19 (2005): 47.
Bonyhady, Tim. *The Colonial Earth*. Melbourne: Miegunyah Press, 2000.
Bowen, Alister M. "'A Power of Money': The Chinese Involvement in Victoria's Early Fishing Industry." PhD diss., La Trobe University, 2008.
Bowman, David. "Bring Elephants to Australia?" *Nature* 482, no. 2 (February 2012): 30.
Boyce, James. *1835: The Founding of Melbourne and the Conquest of Melbourne*. Collingwood: Black Inc., 2011.
———. "Return to Eden: Van Diemen's Land and the Early British Settlement of Australia." *Environment and History* 14, no. 2 (2008): 287–307.
Brandon-Jones, Christine. "Edward Blyth, Charles Darwin, and the Animal Trade in Nineteenth-Century India and Britain." *Journal of the History of Biology* 30, no. 2 (1997): 145–78.
Brennan, Claire. "Imperial Game: A History of Hunting, Society, Exotic Species and the Environment in New Zealand and Victoria, 1840-1901." PhD diss., University of Melbourne, 2004.
Bronstein, Jamie L. *Land Reform and Working-Class Experience in Britain and the United States, 1800-1862*. Stanford, CA: Stanford University Press, 1999.
Broome, Richard. *Aboriginal Victorians: A History since 1800*. Sydney: Allen and Unwin, 2005.
Browne, E. J. *The Secular Ark: Studies in the History of Biogeography*. New Haven, CT: Yale University Press, 1983.
Butcher, Barry W. "Frederick McCoy's Anti-Evolutionism: The Cultural Context of Scientific Belief." *Victorian Naturalist: McCoy Special Issue—Part One* 118, no. 5 (2001): 226–35.
Carruthers, Jane. "Changing Perspectives on Wildlife in Southern Africa, C.1840 to C.1914." *Society and Animals* 13, no. 3 (2005): 183–200.

Cheke, Anthony, and Julian P. Hume. *Lost Land of the Dodo: The Ecological History of Mauritius, Réunion and Rodrigues*. London: Bloomsbury Publishing, 2010.

Chew, Matthew K. "Ending with Elton: Preludes to Invasion Biology." PhD diss., Arizona State University, 2006.

Chisholm, A. H. "Bennett, George (1804-1893)." *Australian Dictionary of Biography Online*. Accessed 28 August 2018. http://adb.anu.edu.au/biography/bennett-george-1770.

Clark, Anna. *The Catch: The Story of Fishing in Australia*. Canberra: NLA Publishing, 2017.

Clark, Brett, and Richard York. "The Restoration of Nature and Biogeography: An Introduction to Alfred Russel Wallace's 'Epping Forest.'" *Organization & Environment* 20, no. 2 (June 2007): 213-34.

Clements, John. *Salmon at the Antipodes: A History and Review of Trout, Salmon and Char and Introduced Coarse Fish in Australasia*. Ballarat: J. Clements, 1988.

Coates, Peter. "Eastenders Go West: English Sparrows, Immigrants, and the Nature of Fear." *Journal of American Studies* 39, no. 3 (2005): 431-62.

Collins, Timothy. "From Anatomy to Zoophagy: A Biographical Note on Frank Buckland." *Journal of the Galway Archaeological and Historical Society* 55 (2003): 91-109.

Colpitts, George. *Game in the Garden: A Human History of Wildlife in Western Canada to 1940*. Vancouver: UBC, 2002.

Coman, Brian J. *Tooth & Nail: The Story of the Rabbit in Australia*. Melbourne: Text Publishing, 1999.

Couzens, E. "Only Half a Penguin a Day: The Early History of Wildlife Law in South Africa." In *The Exemplary Scholar: Essays in Honour of John Milton*, edited by Nathalie J. Chalifour, Patricia KameriMbote, Lin Heng Lye, and John R. Nolo, 207-35. Cape Town: Juta, 2007.

Cowles, Henry M. "A Victorian Extinction: Alfred Newton and the Evolution of Animal Protection." *British Journal for the History of Science* 46, no. 4 (2012): 1-20.

Crosby, Alfred W. *Ecological Imperialism: The Biological Expansion of Europe 900-1900*. Second ed. Cambridge: Cambridge University Press, 2004.

Cushing, D. H. *The Provident Sea*. Cambridge: Cambridge University Press, 1988.

Cushman, Gregory T. *Guano and the Opening of the Pacific World: A Global Ecological History*. New York: Cambridge University Press, 2013.

Dando-Collins, Stephen. *Pasteur's Gambit: Louis Pasteur, the Australasian Rabbit Plague & a Ten Million Dollar Prize*. North Sydney: Vintage Books, 2008.

Darragh, Thomas. "Hall, Thomas Sergeant (1858-1915)." *Australian Dictionary of Biography Online*. Accessed 28 August 2018. http://adb.anu.edu.au/biography/hall-thomas-sergeant-6530.

De Courcy, Catherine. *Evolution of a Zoo: A History of the Melbourne Zoological Gardens, 1857-1900*. Auburn: Quiddlers, 2003.

Diez, Jeffrey M., Jon J. Sullivan, Philip E. Hulme, et al. "Darwin's Naturalization Conundrum: Dissecting Taxonomic Patterns of Species Invasions." *Ecology Letters* 11, no. 7 (2008): 674-81.

Dow, Coral. "A Sportman's Paradise: The Effects of Hunting on the Avifauna of the Gippsland Lakes." *Environment and History* 14, no. 2 (2008): 145-64.

Dow, Gwyneth. "Higinbotham, George (1826–1892)." *Australian Dictionary of Biography Online*. Accessed 28 August 2018. http://adb.anu.edu.au/biography/higinbotham-george-3766.

Dunlap, Thomas R. "Australian Nature, European Culture: Anglo Settlers in Australia." *Environmental History Review* 17, no. 1 (1993): 25–48.

———. *Nature and the English Diaspora: Environment and History in the United States, Canada, Australia and New Zealand*. London: Cambridge University Press, 1999.

———. "Remaking the Land: The Acclimatisation Movement and Anglo Ideas of Nature." *Journal of World History* 8, no. 2 (1997): 203–19.

Dunn, Bob. *Angling in Australia: Its History and Writings*. Balmain: David Ell Press, 1991.

Endersby, Jim. "Escaping Darwin's Shadow." *Journal of the History of Biology* 36 (2003): 358–403.

Finney, Colin. *Paradise Revealed: Natural History in Nineteenth-Century Australia*. Melbourne: Museum of Victoria, 1993.

Fisher, John. "Property Rights in Pheasants: Landlords, Farmers and the Game Laws, 1860–80." *Rural History* 11, no. 2 (2000): 165–80.

Frame, Tom. *Evolution in the Antipodes: Charles Darwin and Australia*. Sydney: University of New South Wales, 2010.

Franklin, A. "Australian Hunting and Angling Sports and the Changing Nature of Human-Animal Relations in Australia." *The Australian and New Zealand Journal of Sociology* 32, no. 3 (1996): 39–56.

———. "Performing Acclimatisation: The Agency of Trout Fishing in Postcolonial Australia." *Ethnos* 76, no. 1 (2011): 19–40.

Frost, Warwick. "Did They Really Hate Trees? Attitudes of Farmers, Tourists and Naturalists towards Nature in the Rainforests of Eastern Australia." *Environment and History* 8, no. 1 (2002): 3–19.

———. "The Environmental Impacts of the Victorian Gold Rushes: Miners' Accounts during the First Five Years." *Australian Economic History Review* 53, no. 1 (2013): 72–90.

———. "European Farming, Australian Pests: Agricultural Settlement and Environmental Disruption in Australia, 1800–1920." *Environment and History* 4, no. 2 (1998): 129–43.

———. "Farmers, Government, and the Environment: The Settlement of Australia's 'Wet Frontier,' 1870–1920." *Australian Economic History Review* 37, no. 1 (1997): 19–38.

Gascoigne, John. "Science and the British Empire from Its Beginnings to 1850." In *Science and Empire: Knowledge and Networks of Science across the British Empire, 1800–1970*, edited by Brett M. Bennett and Joseph M. Hodge, 47–68. New York: Palgrave Macmillan, 2011.

Gess, David, and Sandra Swart. "The Stag of the Eastern Cape: Power, Status and Kudu Hunting in the Albany and Fort Beaufort Districts, 1890 to 1905." *African Historical Review* 46, no. 2 (2014): 48–76.

Gillbank, Linden. "The Acclimatisation Society of Victoria." *Victorian Historical Journal* 51, no. 4 (1980): 255–71.

---. "Animal Acclimatisation: McCoy and the Menagerie That Become Melbourne's Zoo." *The Victorian Naturalist: McCoy Special Issue—Part Two* 118, no. 6 (2001): 297–304.

---. "Of Weeds and Other Introduced Species: Ferdinand Mueller and Plant and Animal Acclimatisation in Colonial Victoria." *The Victorian Naturalist* 124, no. 2 (2007): 69–78.

---. "The Origins of the Acclimatisation Society of Victoria: Practical Science in the Wake of the Gold Rush." *Historical Records of Australian Science* 6, no. 3 (1986): 30–65.

Goodman, David. *Gold Seeking: Victoria and California in the 1850s*. Sydney: Allen and Unwin, 1994.

Grarock, Kate, Christopher R. Tidemann, Jeffrey T. Wood, et al. "Are Invasive Species Drivers of Native Species Decline or Passengers of Habitat Modification? A Case Study of the Impact of the Common Myna (Acridotheres Tristis) on Australian Bird Species." *Austral Ecology* 39, no. 1 (2014): 106–14.

Griffin, Emma. *Blood Sport: Hunting in Britain since 1066*. New Haven, CT: Yale University Press, 2007.

Griffiths, Tom. *Hunters and Collectors: The Antiquarian Imagination in Australia*. Cambridge: Cambridge University Press, 1996.

Grove, Richard. *Ecology, Climate and Empire: Colonialism and Global Environmental History, 1400–1940*. 1st ed. Cambridge: White Horse, 1997.

---. *Green Imperialism: Colonial Expansion, Tropical Island Edens and the Origins of Environmentalism*. Cambridge: Cambridge University Press, 1995.

Hall, Graham P., and Kate P. Gill. "Management of Wild Deer in Australia." *Journal of Wildlife Management* 69, no. 3 (2005): 837–39.

Hall, H. L. "Michie, Sir Archibald (1813–1899)." *Australian Dictionary of Biography Online*. Accessed 28 August 2018. http://adb.anu.edu.au/biography/michie-sir-archibald-4196.

Halverson, Anders. *An Entirely Synthetic Fish: How Rainbow Trout Beguiled America and Overran the World*. New Haven, CT: Yale University Press, 2010.

Harrison, A. J. "The Fisheries Savant: William Saville-Kent in Victoria, 1887–8." *Historical Records of Australian Science* 11, no. 3 (1996): 419–29.

---. *Savant of the Australian Seas: William Saville-Kent (1845–1908) and Australian Fisheries*. Hobart, Australia: Tasmanian Historical Research Association, 1997.

Hoage, R. J., and Deiss, William A., eds. *New Worlds, New Animals: From Menagerie to Zoological Park in the Nineteenth*. Baltimore: Johns Hopkins University Press, 1996.

Hopkins, Harry. *The Long Affray: The Poaching Wars, 1760–1914*. London: Secker and Warburg, 1985.

Houghton, Sheila, and Gary Presland. *Leaves from Our History: The Field Naturalists Club of Victoria*. Box Hill, Australia: Field Naturalists Club of Victoria, 2005.

Humphries, Paul, and Keith F. Walker. "The Ecology of Australian Freshwater Fishes: An Introduction." In *Ecology of Australian Freshwater Fishes*, edited by Paul Humphries and Keith F. Walker, 1–30. Canberra: CSIRO Publishing, 2013.

Hunter, Kathryn M. "New Zealand Hunters in Africa: At the Edges of the Empire of Nature." *Journal of Imperial & Commonwealth History* 40, no. 3 (2012): 483–501.

Hussain, Shafqat. "Sports-Hunting, Fairness and Colonial Identity: Collaboration and Subversion in the Northwestern Frontier Region of the British Indian Empire." *Conservation & Society* 8, no. 2 (2010): 112–26.

Hutton, Drew, and Libby Connors. *A History of the Australian Environment Movement*. Melbourne: Cambridge University Press, 1999.

Ito, Takashi. *London Zoo and the Victorians, 1828-1859*. New York: Boydell and Brewer Group, 2014.

Jacobsen, Lif Lund. "Steam Trawling on the South-East Continental Shelf of Australia: An Environmental History of Fishing, Management and Science in New South Wales, 1865-1961." PhD diss., University of Tasmania, 2010.

Jeffries, Stephen. "Alexander Von Humboldt and Ferdinand Von Mueller's Argument for the Scientific Botanic Garden." *Historical Records of Australian Science* 11, no. 2 (1996): 301–10.

Jones, P. A., and Anna Kenny. *Australia's Muslim Cameleers: Pioneers of the Inland, 1860s-1930s*. Adelaide: Wakefield Press, 2010.

Judd, Richard William. *Common Lands, Common People: The Origins of Conservation in Northern New England*. Cambridge, MA: Harvard University Press, 1997.

———. *The Untilled Garden: Natural History and the Spirit of Conservation in America, 1740-1840*. New York: Cambridge University Press, 2009.

Karsten, Peter. *Between Law and Custom: High and Low Legal Cultures in the Lands of the British Diaspora the United States, Canada, Australia, and New Zealand, 1600-1900*. New York: Cambridge University Press, 2002.

Kenny, Robert. *The Lamb Enters the Dreaming: Nathanael Pepper and the Ruptured World*. Carlton North, Australia: Scribe Publications, 2007.

Kinsey, Darin. "'Seeding the Water as the Earth': The Epicenter and Peripheries of a Western Aquacultural Revolution." *Environmental History* 11, no. 3 (2006): 527–66.

Kurlansky, Mark. *The Big Oyster: History on the Half Shell*. 1st ed. New York: Ballantine Books, 2006.

Lambert, David, and Alan Lester. "Imperial Spaces, Imperial Subjects." In *Colonial Lives across the British Empire: Imperial Careering in the Long Nineteenth Century*, edited by David Lambert and Alan Lester, 1–31. Cambridge: Cambridge University Press, 2006.

Landry, Donna. *The Invention of the Countryside: Hunting, Walking and Ecology in English Literature, 1671-1831*. Basingstoke: Palgrave, 2001.

Leslie, D. J. "Moira Lake: A Case Study of the Deterioration of a River Murray Natural Resource." PhD diss., University of Melbourne, 1995.

Le Souëf, J. Cecil. "The Development of a Zoological Garden at Royal Park." *Victorian Historical Magazine* 37, no. 4 (1966): 221–52.

Lester, Alan. "Imperial Circuits and Networks: Geographies of the British Empire." *History Compass* 4, no. 1 (2006): 124–41.

Levasseur, Olivier, and Darin Kinsey. "The Second Empire Legacy of the French 'Culture' of Oysters." *International Journal of Maritime History* 20, no. 1 (2008): 253–68.

Lever, Christopher. *Naturalized Fishes of the World*. San Diego, CA: Academic Press, 1996.

———. *They Dined on Eland: The Story of the Acclimatisation Societies*. London: Quiller, 1992.

Lewis, Annette. "The Life and Times of Peter Snodgrass 1817–1867." *Victorian Historical Journal* 81, no. 2 (2010): 214–33.

Lucas, A. M. "Ferdinand Von Mueller's Interactions with Charles Darwin and His Response to Darwinism." *Archives of Natural History* 37, no. 1 (2010): 103–30.

Lucas, A. M., Sara Maroske, and Andrew Brown-May. "Bringing Science to the Public: Ferdinand Von Mueller and Botanical Education in Victorian Victoria." *Annals of Science* 63, no. 1 (2006): 25–57.

Macintyre, Stuart Forbes, and Richard Joseph W. Selleck. *A Short History of the University of Melbourne*. Carlton: Melbourne University Press, 2003.

MacKenzie, John M. *The Empire of Nature: Hunting, Conservation and British Imperialism*. Manchester, UK: Manchester University Press, 1988.

MacLeod, Roy M. *Archibald Liversidge, F.R.S.: Imperial Science under the Southern Cross*. Sydney: Sydney University Press, 2009.

———. "Government and Resource Conservation: The Salmon Acts Administration, 1860–1886." *Journal of British Studies* 7, no. 2 (1968): 114–50.

Maroske, Sara. "Science by Correspondence: Ferdinand Mueller and Botany in Nineteenth Century Australia." PhD diss., University of Melbourne, 2005.

Marshall, A. J. "The World of Hopkins Sibthorpe." In *The Great Extermination*, edited by A. J. Marshall, 9–18. Melbourne: Panther Press, 1965.

Mason, Peter. *Before Disenchantment: Images of Exotic Animals and Plants in the Early Modern World*. London: Reaktion Books, 2009.

McCalman, Iain. *The Reef: A Passionate History*. Melbourne: Viking Press, 2013.

McCann, Doug. "Frederick McCoy and the Naturalist Tradition." *Victorian Naturalist: McCoy Special Issue—Part Two* 118, no. 6 (2001): 209–14.

———. "Timeline: Frederick McCoy." *Victorian Naturalist: McCoy Special Issue—Part One* 118, no. 5 (2001): 148–50.

McEvey, Allan. "Le Souef, Albert Alexander (1828–1902)." *Australian Dictionary of Biography Online*. Accessed 28 August 2018. http://adb.anu.edu.au/biography/le-souef-albert-alexander-4013.

McGowan, Barry. "Mullock Heaps and Trailing Mounds: Environmental Effects of Alluvial Goldmining." In *Gold: Forgotten Histories and Lost Objects of Australia*, edited by Iain McCalman, Alexander Cook, and Andrew Reeves, 85–103. Cambridge: Cambridge University Press, 2001.

McKenzie, Kirsten. *Scandal in the Colonies: Sydney & Cape Town, 1820–1850*. Carlton: Melbourne University Press, 2004.

McPeek, Mark A. "Limiting Factors, Competitive Exclusion, and a More Expansive View of Species Coexistence." *American Naturalist* 183, no. 3 (2014): iii–iv.

Mennell, Philip. *The Dictionary of Australasian Biography: Comprising Notices of Eminent Colonists from the Inauguration of Responsible Government Down to the Present Time (1855–1892)*. London: Hutchinson, 1892.

Mickle, Alan. D. "Lyall, William (1821–1888)." *Australian Dictionary of Biography Online*. Accessed 28 August 2018. http://adb.anu.edu.au/biography/lyall-william-4051.

Minard, Peter. "Assembling Acclimatization: Frederick McCoy, European Ideas, Australian Circumstances." *Historical Records of Australian Science* 24, no. 1 (2013): 1–14.
———. "A History of Zoological Acclimatisation in Victoria, 1858–1900." PhD diss., University of Melbourne, 2015.
———. "Salmonid Acclimatisation in Colonial Victoria: Improvement, Restoration and Recreation 1858–1909." *Environment and History* 21, no. 2 (2015): 177–99.
Mitchell, Jessie. "Alpacas in Colonial Australia: Acclimatisation, Evolution and Empire." *Journal of Australian Colonial History* 12 (2010): 55–76.
Moore-Colyer, R. J. "Feathered Women and Persecuted Birds: The Struggle against the Plumage Trade, C. 1860–1922." *Rural History* 11, no. 1 (2000): 57–73.
Moyal, Ann. *A Bright & Savage Land: The Science of a New Continent—Australia—Where All Things Were 'Queer and Opposite.'* Melbourne: Penguin Books, 1986.
Mulvaney, D. J. "Spencer, Baldwin (1860–1929)." *Australian Dictionary of Biography Online*. Accessed 28 August 2018. http://adb.anu.edu.au/biography/spencer-sir-walter-baldwin-8606.
Mulvaney, D. J., and J. H. Calaby. *"So Much That Is New": Baldwin Spencer, 1860–1929, a Biography*. Carlton: Melbourne University Press, 1985.
Nash, Colin E. *The History of Aquaculture*. Ames, IA: Wiley-Blackwell, 2011.
Newland, Elizabeth Dalton. "Dr George Bennett and Sir Richard Owen: A Case Study of the Colonization of Early Australian Science." In *International Science and National Scientific Identity: Australia between Britain and America*, edited by R. W. Home and Sally Gregory Kohlstedt, 55–75. London: Kluwer Academic Publishers, 1991.
Noble, J. C., and G. H. Pfitzner. "'They Know Not What They Do'; on William Rodier and His Mission to Exterminate Rabbits and Other Vertebrate Pests." *Historical Records of Australian Science* 14, no. 4 (2002): 431–57.
Norman, F. I., and A. D. Young. "Short-Sighted and Doubly Short-Sighted Were They: A Brief Examination of the Game Laws of Victoria." *Journal of Australian Studies* 7, no. 3 (1980): 5–24.
Norrie, Justin. "Elephants the Answer to Bushfire Problem? That's Dumbo, Scientists Say." *The Conversation Media Group*. Accessed 28 August 2018. http://theconversation.com/elephants-the-answer-to-bushfire-problem-thats-dumbo-scientists-say-5144.
Olsen, Penny. *Upside Down World: Early European Impressions of Australia's Curious Animals*. Canberra: National Library of Australia, 2010.
Osborne, Michael A. "Acclimatising the World: A History of a Paradigmatic Colonial Science." *Osiris* 15, no. 2 (2000): 135–51.
———. *Nature, the Exotic, and the Science of French Colonialism*. Bloomington: Indiana University Press, 1994.
Ostapenko, Dmytro. "Golden Horizons: Expansion of the Wheat-Growing Industry in the Colony of Victoria in the 1850s." *Agricultural History* 87, no. 1 (2013): 35–56.
Petrow, Stefan. "Civilizing Mission: Animal Protection in Hobart 1878–1914." *Britain & the World* 5, no. 1 (2012): 69–95.
Philipp, June. *A Great View of Things: Edward Gibbon Wakefield*. Melbourne: Thomas Nelson and Sons, 1971.

Pittie, A. *Birds in Books: Three Hundred Years of South Asian Ornithology: A Bibliography*. Calcutta: Permanent Black, 2010.
Powell, J. M. *Environmental Management in Australia, 1788-1914, Guardians, Improvers and Profit: An Introductory Survey*. Melbourne: Oxford University Press, 1976.
———. *Watering the Garden State: Water, Land and Community in Victoria 1834-1988*. Sydney: Allen and Unwin, 1989.
Presland, Gary. *The Place for a Village: How Nature Has Shaped the City of Melbourne*. Melbourne: Museums Victoria Publishing, 2008.
Reynolds, Henry. *The Law of the Land*. 2nd ed. Ringwood, Australia: Penguin, 1992.
Rideout, Roger. "'Handsome Gifts' to a Young Society." *History Today* 62, no. 1 (2012): 37–43.
Ritvo, H. "Going Forth and Multiplying: Animal Acclimatization and Invasion." *Environmental History* 17, no. 2 (2012): 404–14.
Roberts, Stephen H. *History of Australian Land Settlement 1788-1920*. Melbourne: MacMillan of Australia, 1968.
Robin, Libby. *How a Continent Created a Nation*. Sydney: University of New South Wales Press, 2007.
Robinson, Robb. "The Evolution of Some Key Elements of British Fisheries Policy." *International Journal of Maritime History* 9, no. 2 (1997): 129–50.
Roche, Chris. "'Fighting Their Battles O'er Again': The Springbok Hunt in Graaff-Reinet, 1860–1908." *Kronos* 29 (2003): 86–108.
Rolls, Eric C. *More a New Planet than a New Continent*. Canberra: Australian National University, 1985.
———. *They All Ran Wild: The Animals and Plants That Plague Australia*. 2nd ed. Melbourne: Angus and Robertson, 1984.
Ronald, Heather B. *Hounds Are Running: A History of the Melbourne Hunt*. Kilmore: Lowden, 1970.
Rookmaaker, Kees. "The Zoological Contributions of Andrew Smith (1797–1872) with an Annotated Bibliography and a Numerical Analysis of Newly Described Animal Species." *Transactions of the Royal Society of South Africa* 72, no. 2 (2017): 105–73.
———. *The Zoological Exploration of Southern Africa, 1650-1790*. Rotterdam: A. A. Balkema, 1989.
Rowland, S. J. "Aspects of the History and Fishery of the Murray Cod, Maccullochella Peeli." *Proceedings of the Linnean Society of New South Wales* 111, no. 3 (1989): 210–13.
———. *Overview of the History, Fishery, Biology and Aquaculture of Murray Cod (Maccullochella Peelii Peelii)*. Sydney: NSW Department of Primary Industries, 2004.
Rupke, Nicolaas. *Richard Owen: Biology without Darwin, a Revised Edition*. 2nd ed. London: University of Chicago Press, 2009.
Saunders, Brian. *The Discovery of Australian Fishes: A History of Australian Ichthyology to 1930*. Collingwood: CSIRO Publishing, 2012.
Serle, Geoffery. *The Golden Age: A History of the Colony of Victoria, 1851-1861*. 2nd ed. Melbourne: Melbourne University Press, 1968.

---. "Wilson, Edward (1813-1878)."Australian Dictionary of Biography Online. Accessed 28 August 2018. http://adb.anu.edu.au/biography/wilson-edward-4866.
Serventy, D. L. "The Menace of Acclimatisation." *Emu: The Journal of the Australasian Ornithological Society* 36, no. 1 (1937): 189-97.
Shiel, D. J. *Charles Jardine Don, Australian Labour's First Parliamentarian: The People's Man*. North Melbourne: Garravembi, 1996.
Skead, C. J. *Historical Mammal Incidence in the Cape Province*. Cape Town: Department of Nature and Environmental Conservation of the Provincial Administration of the Cape of Good Hope, 1980.
Smith, Neil. "Sir James Arndell Youl (1811-1904)." Australian Dictionary of Biography Online. Accessed 28 August 2018. http://adb.anu.edu.au/biography/youl-sir-james-arndell-4899/text8199.
Star, Paul. "From Acclimatisation to Preservation: Colonists and the Natural World in Southern New Zealand 1860-1894." PhD diss., University of Otago, 1997.
---. "Plants, Birds and Displacement Theory in New Zealand, 1840-1900." *British Review of New Zealand Studies* 10 (1997): 5-21.
Stockland, Etienne. "Policing the Oeconomy of Nature: The Oiseau Martin as an Instrument of Oeconomic Management in the Eighteenth-Century French Maritime World." *History & Technology* 30, no.3 (2014): 207-31.
Stubbs, Brett J. "From 'Useless Brutes' to National Treasures: A Century of Evolving Attitudes Towards Native Fauna in New South Wales, 1860s to 1960s." *Environment and History* 7, no. 1 (2001): 23-56.
Taylor, Joseph E. *Making Salmon: An Environmental History of the Northwest Fisheries Crisis*. Seattle, WA: University of Washington Press, 1999.
Thompson, E. P. *Whigs and Hunters: The Origin of the Black Act*. 1st American ed. New York: Pantheon Books, 1975.
Thuiller, Willifred, Laure Gallien, Isabelle Boulangeat, et al. "Darwin's Naturalization Conundrum: A Quest for Evidence." *Diversity and Distributions* 16, no. 3 (2010): 461-75.
Towle, Terry C. "Authored Ecosystems: Livingstone Stone and the Transformation of Californian Fisheries." *Environmental History* 5, no. 1 (2000): 54-74.
Tyrrell, Ian. "Acclimatisation and Environmental Renovation: Australian Perspectives on George Perkins Marsh." *Environment and History* 10, no. 2 (2004): 153-67.
---. *True Gardens of the Gods: Californian Australian Environmental Reform 1860-1930*. Los Angeles: University of California Press, 1999.
University of British Columbia Fisheries Centre. "Fishbase." Fishbase.org. Accessed 28 August 2018. http://www.fishbase.org/Nomenclature/ValidNameList.php?syng=salmo&syns=&vtitle=Scientific+Names+where+Genus+Equals+%3Ci%3E Salmo%3C%2Fi%3E&crit2=CONTAINS&crit1=EQUAL.
Van Sittert, Lance. "Bringing in the Wild: The Commodification of Wild Animals in the Cape Colony/Province C. 1850-1950." *Journal of African History* 46, no. 2 (2005): 269-91.
Walker, Jean. *Origins of the Tasmanian Trout: An Account of the Salmon Ponds and the First Introduction of Salmon and Trout to Tasmania in 1864*. Hobart, Australia: Inland Fisheries Commission, 1988.

Waterhouse, Richard. *The Vision Splendid: A Social and Cultural History of Rural Australia*. Fremantle, Australia: Curtin University Books, 2005.

Weiner, Douglas R. "The Roots of 'Michurinism': Transformist Biology and Acclimatization as Currents in the Russian Life Sciences." *Annals of Science* 42, no. 3 (1985): 243–60.

Wells, Philippa. "'An Enemy of the Rabbit': The Social Context of Acclimatisation of an Immigrant Killer." *Environment and History* 12, no. 3 (2006): 297–324.

Wilkins, Noel P. *Ponds, Passes, and Parcs: Aquaculture in Victorian Ireland*. Dublin: Glendale, 1989.

Index

Aboriginal, 3–4, 32, 38, 58, 62–63, 89, 109, 113, 151n12. *See also* Wurundjeri; Yorta Yorta
Acclimatisation Society of New South Wales, 24, 63, 166n1
Acclimatisation Society of Victoria: formation, 7–17, 150n6; decline, 108–21; theories, 2, 23–29, 37–38, 112–14
Africa, 13, 15, 28–31, 34, 37, 39–51, 55, 80, 86, 142–46.
Agassiz, Louis, 26, 28, 41, 155n20.
Algeria, 2, 139
Allport, Morton, 79–80, 82
alpacas, 1, 9, 12–13, 18–19, 68, 86, 108, 120
angling, 58, 60–61, 65–68, 74, 83, 121–24, 163n1, 174n36
antelope, 29–31, 43, 47, 49–51, 95, 97–98, 108, 141, 144, 147
aquaculture, 15, 36; origins of, 73–75; Australian development, 76–80, 121–28, 163n1
Ashworth, Edward, 74, 76, 78
Ashworth, Thomas, 74, 76, 78
axis deer (*Axis axis*), 53–54, 70, 85, 140, 147

balance of nature, 23–24, 29, 31–37, 40–41, 113–14, 117–18, 132, 136
Ballarat, 10, 70, 92, 124–25, 128
Ballarat Fish Acclimatizing Society, 70, 124–25
bandicoot, 31, 139
Barkly, Henry, 16, 37, 43, 45
Bendigo, 58, 63, 99, 112
Bennett, George, 12, 23–24, 29–32, 37, 40–42, 44, 47, 55, 64, 96, 134, 156n38–39

Berkeley, Grantley, 14
Bindon, Samuel, 61, 102
Black, Alexander, 75–77, 165n22
Black, Thomas, 17, 19, 21, 67–69, 80, 104, 116, 117, 152n8
blackfish, (*Gadopsis marmoratus*), 58, 62, 121, 130–32
black swans (*Cygnus atratus*), 15, 44, 48, 55, 92, 95, 139
Blyth, Edward, 52–54, 147, 160n53
Boccius, Gottlieb, 73, 75
Bright, John, 88, 94
Britain, 1–8, 11–12, 45–46, 55–57, 62, 66, 73–75, 96–98, 127–28.
British Empire, 1, 5–6, 12, 14, 21, 36–37, 72, 86, 121, 127 135
brolgas (*Antigone rubicunda*), 97, 168n67
bronze-winged pigeons (*Phaps chalcoptera*), 95–96
brush turkey (*Alectura lathami*), 15, 31–32, 97, 139
Buckland, Frank, 13–16, 43–44, 47, 76, 78, 116, 118–19, 125, 172n68
Burnett, James Ludovick, 75
Butler, Edward, 52, 54

Campbell, Archibald, 111, 114–16, 118, 171n50
Canada, 86, 107
Cape Barren geese (*Cereopsis novaehollandiae*), 44, 48, 140
Cape Colony, 43, 47–51, 55, 141
carp (*Cyprinus carpio*), 75–76, 142, 147
Ceylon, 2, 11, 30, 31, 48, 51–55, 85, 139
Chile, 1, 18
China, 30–31, 40, 58, 141
Chirnside, Andrew, 98
Coste, Victor, 73–75

193

Crockford, John, 13
Cuvier, Georges, 64–65, 162n54

Darwin, Charles, 18, 25, 27–28, 37–38, 41, 111–12, 116, 135. *See also* evolution
deer, 28–29, 36, 43, 46, 53–55, 86–89, 95, 98, 104, 129. *See also* axis deer; fallow deer; hog deer; sambar deer; red deer
dingo, 25, 38, 88–90, 95–96, 119
displacement theory, 113, 117–19, 131, 154n1, 172n68
Don, Charles, 93, 167n50
Duffield, James, 18–19
Duffy, Charles Gavan, 104–5, 167n50, 168n82

Echuca, 63, 105
ecological imperialism, 4–5, 10, 21, 29, 31, 35, 38, 42, 73, 134, 152n19; neo-ecological imperialism, 4–5, 31, 35, 38, 42, 73, 134
eland (*Taurotragus oryx*), 2, 13, 31, 43, 47, 51, 141, 145
Embling, Thomas, 9, 18
emus (*Dromaius novaehollandiae*), 1, 32, 43–45, 55, 88, 90, 95
Espeut, William Bancroft, 119
evolution, 3, 23–26, 37, 41, 111–13, 135, 154n1, 157n73, 170n24
extinction, 111, 113, 115–16, 118, 126, 170n24

fallow deer (*Dama dama*), 48, 90, 98
Fence, Field, and Chattel Preservation League, 104
Field Naturalists Club of Victoria, 106, 110, 112–16, 131
Fiji, 2
fishing: commercial, 58–61; Chinese, 58–59, 68–69; regulation, 59–60, 66–70, 78
Forbes, Edward, 26, 28, 41, 155n20
France, 2, 5 45, 55, 62, 63, 66, 73–74, 77, 139–40, 142, 154n1, 172n87

Francis, George, 12
French, Charles, 111
Fyans, Foster, 90

galaxia, 64–65, 124
gazelles, 47, 141
Geelong, 10, 58, 69, 96, 102, 123–24, 173n9
Geelong and Western District Fish Acclimatising Society, 70, 123–24
Geoffroy Saint-Hilaire, Isidore, 12, 14, 44, 153n23
golden perch (*Macquaria ambigua*), 58–60, 124, 140
gold rush, 8–9, 35, 57–59, 71, 76, 89, 91, 101, 106–7, 151n13
Gould, John, 15, 25, 31, 41, 44
gourami, 2, 43, 142, 151n7
grayling, Australian (*Prototroctes maraena*), 58, 64–65, 78–79, 125–26, 128
Grey, Sir George, 48
Günther, Albert, 64–65, 125, 162n54, 165n38

Haines, William, 21, 100–101
Hares, 1–2, 38–40, 46–47, 97–99, 102–10, 117, 120, 129, 140–43, 147; spring hares, 47, 140
Higinbotham, George, 103, 168–69n100
hog deer (*Hyelaphus porcinus*), 1, 53–54, 140, 147
Horticultural Society of Victoria, 108
Humboldt, Alexander von, 10, 28, 32, 33–34, 37, 42, 153n15, 165n15
hunting, 85–108; English traditions, 87–89; clubs, 90; commercial, 88, 90–91, 101, 105; regulation of, 91–103, 114–16, 166n1
Hutton, Frederick W., 118

India, 29–31, 34, 37, 43, 45–46, 51–58, 86, 107, 135, 139
Indian mynas (*Acridotheres tristis*), 1–4, 43, 54–55, 117, 129, 141, 160n69

194   *Index*

Jones, William, 93

kangaroos (*Macropus giganteus*), 1, 20, 31–32, 38–40, 44–45, 55, 88–91, 95–96, 119, 139
kookaburras (*Dacelo novaeguineae*), 1, 15, 20, 31–32, 44–45, 96–97, 139
Krefft, Gerard, 64–65

Landells, George James, 52
land reform, 8–11, 19, 21–22, 42, 89, 92–93, 100, 107, 129, 151n13, 168n82
Lang, Samuel, 59–60
Layard, Charles, 48, 52, 54
Layard, Edgar Leopold, 48–52, 143
Ledger, James, 18–19
Le Souëf, Albert, 109–17, 122, 169n1
Le Souëf, Dudley, 109–11, 120, 122, 130, 133
Lewis, Frederick, 131
lyrebirds (*Menura novaehollandiae*), 32, 115

Macdonald, Donald, 121, 130–32
Mackinnon, Lachlan, 79, 94
Madden, Henry Ridgewood, 23–24, 37–38, 40–43, 80, 114, 134
magpies (*Gymnorhina tibicen*), 1, 20, 40–45, 11, 139
Maribyrnong River, 68
Marsh, George Perkins, 33, 36–37, 41, 72, 75, 82, 157n52
Mauritius, 2, 30, 43, 45, 139–40, 142, 160n69
McCoy, Frederick, 12, 18, 232–4, 26–30, 34, 37, 39, 40–44, 47, 53, 56, 60–62, 64–65, 67, 69, 79–80, 110–11, 113–14, 134–36, 152n8, 155n20
Meek, James, 68
Melbourne Hunt Club, 90, 98–99, 103, 106
Michie, Archibald, 103, 168n100
Mitchell, David, 13
mongoose, 118–19
Morton, William Lockhart, 61

Mueller, Ferdinand von, 18, 23–25, 32–37, 41–44, 53–55, 62, 111, 134–35, 151n8, 157n52
Mullick, Rajendra, 52, 54
murray cod (*Maccullochella peelii*), 10, 15, 58, 61–63, 124, 126, 140
Murray River Fishing Company, 58, 63

New South Wales, 3, 7–8, 12, 31, 37, 41, 59, 62–63, 98
Newton, Alfred, 111, 118–19, 170n24
New Zealand, 1–2, 48–49, 112–13, 118–19, 123–25, 130, 139–40, 143–46, 154n1, 172n87
Nicholson, Charles, 67

ostriches (*Struthio camelus*), 43, 47, 50, 109, 142
Owen, Richard, 13–14, 24–26, 28, 30, 39–41, 114, 155n20
oysters, 58, 60, 68, 102

partridges, 20, 31, 50–51, 53–55, 95, 120, 146–47
perch, English (*Perca fluviatilis*), 78, 122, 124
pheasants, 20, 31, 48, 50–51, 54–55, 95–99, 110, 120, 145–47, 161n73
Phillip Island, 39–40, 55, 58, 104, 161n73
Philosophical Institute of Victoria, 9–10, 12, 25, 33, 152n8
Port Phillip Bay, 3, 18, 58–60, 67–68
possums, 28, 30–31, 38
Putwain, James, 60, 66–67

quolls, 31, 39, 116

rabbits, (*Oryctolagus cuniculus*), 4, 39–40, 46, 87–88, 96–98, 104, 110, 113, 116–21, 172n68
Ramsbottom, Robert, 74–75, 78
red deer (*Cervus elaphus*), 20, 46, 98, 118, 141
Royal Botanical Gardens, 2, 10, 34, 153n8

Index 195

Royal Melbourne Zoological Gardens, 1, 109, 130, 133, 169n1
Royal Park, 1-2, 39-40, 47, 51, 53-55, 82, 117, 122, 130, 133.

salmon, 12-13, 15, 60, 62, 64, 81, 163n1; Atlantic salmon (*Salmo salar*), 73-75, 77-80, 82-83, 125, 128, 142; Californian salmon (*Oncorhynchus tshawytscha*), 125, 128, 173n18
sambar deer (*Rusa unicolor*), 1, 53-54, 147
Saville-Kent, William, 127-31
sea trout, (*Arripis trutta*), 79
Selwyn, Alfred, 20, 24-25
Sicily, 2, 139
Smith, Louis Lawrence, 102
Snodgrass, Peter, 93, 167n48
Société Zoologique d'Acclimatation, 1, 12-13, 17, 20-21, 45, 47, 73, 125, 154n1
Society for the Acclimatisation of Animals, Birds, Insects and Vegetables within the United Kingdom, 1, 13-16, 20, 43, 118, 154n1, 172n68
South America, 18-19, 29-30, 34, 37, 41-42, 46, 48, 65, 135, 139-40
South Australia, 11-12, 48
sparrows (*Passer domesticus*), 1-4, 17, 20, 38, 40-41, 50, 54; as pests, 108-10, 119-21, 129, 134, 158n100
Spencer, Baldwin, 111-14, 118, 135
Sprigg, George, 66, 82, 98
springbok (*Antidorcas marsupialis*), 2, 47, 51, 144. *See also* antelope
Stone, Livingstone, 72, 124-25
swivel guns, 101-3, 105-6

Tasmania, 3, 12-13, 15, 72-83, 122-24, 126-28, 131, 165n38
taxonomy, 26-29, 56, 60-61, 64-65, 126
trout, 17, 43, 65, 73-76, 80-83, 108-9, 120, 122-27, 129-32, 142; brown trout (*Salmo trutta*), 80-83, 122-31, 142,163n1, 164n5; rainbow trout (*Oncorhynchus mykiss*), 124, 130

United States Fisheries Commission, 122-24, 130
University of Melbourne, 2, 12, 18, 23-24, 111

Valenciennes, Achille, 64-65
Victorian Angling Society, 59-68, 115, 126

Wakefield, Edward, 11
Wallace, Alfred Russell, 111-15, 118, 154n1, 170n24
Watson, George, 90, 98
Watts, Henry Edwards, 52, 61, 65
Webber, William, 69
Were, Johnathan Binn, 108
Western Port, 58-61, 67-68, 91, 102
Westgarth, William, 89
Wheelwright, William, 65, 90-91, 96-97
Wilson, Edward, 5, 7-21, 61-62, 75-76, 98-99, 110, 117, 121, 134, 152n4
Wilson, Samuel, 70, 122, 125-26
Wilson's Promontory National Park, 114
wombat (*Vombatus ursinus*), 15, 20, 31
Wonga Pigeon (*Leucosarcia melanoleuca*), 15, 31, 139
Wurundjeri, 3. *See also* Aboriginal

Yarra River, 10, 59, 80, 89, 126
Yorta Yorta, 58-59, 63. *See also* Aboriginal
Youl, James, 74-76

Zebra, 51, 143
Zoological and Acclimatisation Society of Victoria, 83, 105-6, 109-10, 112-15, 122-23
Zoological Society of London, 13, 20, 30-31, 51-52, 119-12, 114-20, 122-27, 130, 140

Made in United States
Orlando, FL
29 July 2023